DO WE NEED MIDWIVES?

DO WE NEED MIDWIVES?

MICHEL ODENT

pinter
&
martin

Do we need midwives?

First published in Great Britain by Pinter & Martin Ltd 2015, reprinted 2016

© 2015 Michel Odent

Michel Odent has asserted his moral right to be identified as the author of this work in accordance with the Copyright, Designs and Patents Act of 1988.

ISBN 978-1-78066-220-6

Cover painting, Young Woman Knitting, by Vincent van Gogh

British Library Cataloguing-in-Publication Data
A catalogue record for this book is available from the British Library.

Set in Minion

Printed and bound in the UK by Ashford Colour Press Ltd, Gosport, Hampshire

This book has been printed on paper that is sourced and harvested from sustainable forests and is FSC accredited.

Pinter & Martin Ltd
6 Effra Parade
London SW2 1PS
www.pinterandmartin.com

CONTENTS

1   A ludicrous question   9

2   A sensible question   12
Pre-midwifery societies   13
Biased knowledge   17
The specifically human handicap   19

3   A useless question   23
Towards a new understanding of normality   23
Preliminary signs of a new normality   28
Pre-labour versus in-labour caesareans   29
What if…?   33

4   primalhealthresearch.com versus NIH   35
From knowledge to awareness   35
A useful tool   36
Concordant results   39
Genesis of non-communicable diseases   41
Enlarging our horizon   43
Limits to primal health research   44

5   The driving force   47
Epigenetics   47
Twenty-first-century bacteriology   49
Informational substances and their receptors   53
Evolutionary biology   56

6	Bridges between scientific perspectives	59
	Autism	59
	Obesity	64
	Interdisciplinarity	69
7	Meanwhile	70
	Timing of the operation	70
	Immediate skin-to-skin contact	71
	Entering the world of microbes	72
	Ironies	75
8	*Homo Ludens* from a primal health research perspective	77
	From Plato to Kerstin Uvnäs Moberg	78
	The primal health research perspective	79
9	At the edge of the precipice	83
	A twentieth-century scientific discovery	83
	Immediate practical implications	86
	A transitory phase of our history?	87
10	The gaps between science and tradition	88
	Countless examples	88
	An unexpected way to learn about the crucial divergence	90
	Another basic physiological concept	94
	The still dominant paradigm	95
11	In pain thou shalt bring forth children	98
12	Will the symbiotic revolution take place?	104
	'Symbiosis' as the antithesis of 'Domination'	104
	Reinventing Fire... and Birth	106
	Flirting with utopia	109
13	What is the sex of angels?	111
	Lesson from an analogy	112
	After the paradigm shift	113

Addendum (To be read after July 2030):
  Can humanity survive medicine?   116
A premature question   116
Neutralised laws of natural selection   117
A vicious circle   119
Genetically modified human beings   120
Different orders of magnitude   121

References   123

Index   143

Special thanks to Liliana Lammers
She gave the title of this book... then I wrote the text.

CHAPTER ONE

A ludicrous question

Midwifery is among the oldest professions. It is one of the characteristics that set our species apart from other mammals: human females routinely seek assistance when they give birth. This is well known among primatologists, as well as in anthropological and medical circles. The specifically human mechanical difficulties are the bases for all sorts of interpretations. Countless articles and textbooks have reproduced the classical drawings by Adolph Schultz.[1,2,3] These drawings show the size of the neonatal skull in relation to the maternal pelvic outlet among spider monkeys, proboscis monkeys, macaques, gibbons, orang-utans, gorillas, common chimpanzees and *Homo sapiens.*

The well-accepted conclusions of the evolutionary perspective are easy to summarise: the socialisation of birth is an adaptation to mechanical difficulties. This is, in part, a result of the fact that the human infant emerges from the birth canal facing the opposite direction from the mother, hindering her ability to assist in its delivery. The presence of another individual who can receive the infant during delivery is, therefore, quasi-necessary. It is claimed that, as bipedalism evolved, natural selection favoured the behaviour of seeking assistance during birth. Seeking companionship was driven by fear, anxiety, pain and desire to conform to behavioural norms. It is also commonplace to claim that the evolutionary process has resulted in heightened emotional needs during labour, which leads women to seek companionship at this time: this suggests that the desire for supportive, familiar people at birth is deeply rooted in human evolutionary history.

The interpretations offered by comparative anatomy are supported by anthropological studies. It would take volumes to review all the perinatal beliefs and rituals reported in a great variety of cultures. As early as 1884, *Labor Among Primitive Peoples*, by George Engelmann, provided an impressive catalogue of beliefs and rituals occurring in hundreds of ethnic groups on all continents.[4] It seems acceptable to conclude from such studies that the socialisation of childbirth is universal.

The academic perspective is in perfect agreement with our cultural conditioning: we believe that a woman does not have the power to give birth by herself. Looking at the roots of the words is a way to illustrate our deep-rooted way of thinking. For example, the origin of the word obstetrics is 'obstetrix' ('midwife' in Latin), which literally means 'the woman staying in front of' (from the verb *'obstare'*). It implies that when a woman is giving birth somebody must be in front of her. Looking at how we talk about childbirth in daily conversations is another way to illustrate the dominant way of thinking. If we ask a young mother who delivered her baby, we are not expecting the answer: 'I gave birth'.

Vocabulary that emerged during the second half of the twentieth century has reinforced our cultural conditioning. Among groups promoting 'natural childbirth' certain terms became popular. A 'coach' is a guide bringing her (his) expertise. The need for 'emotional support' implies that to give birth a woman needs some energy brought by somebody else. In the medical literature the term 'labour management' is widely used. The terms 'coaching' and 'managing' express the same way of thinking.

Since there is a perfect agreement between academics, tradition and contemporary culture, it is obviously ludicrous to raise a question such as: 'Do we need midwives?'

CHAPTER 2

A sensible question

Will our question remain ludicrous after looking critically at these two common assumptions? We'll first ask whether the active participation of a birth attendant is really a universal human characteristic. Then we'll consider whether morphological and mechanical factors are really the main reasons for difficult births in our species.

When studying inherent human nature, we should always take into account the spectacular turning point in the history of mankind that started about 10,000 years ago. Before that time human beings took advantage of what nature could offer, obtaining their food from wild plants and animals. An increased level of social tolerance was an evolutionary advantage. It is notable that *Homo sapiens* gradually developed characteristics suggestive of a process of 'self-domestication', such as facial diminution (a decrease in size and prominence of the facial features). Then, suddenly, our ancestors began to domesticate plants and animals in places as diverse and as far apart from each other as the Tigris and Euphrates valleys in the Middle East, South Asia, Central Asia, and Central America. The advent of agriculture and animal husbandry – the

Neolithic revolution – radically changed the dominant human lifestyle. Our ancestors were obliged to be less nomadic and more sedentary. The concept of territory took on unprecedented importance, along with other reasons for conflicts between human groups. The new basic strategy for survival was to dominate nature and dominate other human groups. It became an advantage to develop the huge human potential for aggression.

It is acceptable – although simplistic – to claim that there have been two eras in the history of *Homo sapiens*, separated by the Neolithic revolution. Yet today even the most authoritative academics may be imprisoned by their own conditioning, which has gradually developed during the comparatively short period following this revolution. The way human babies are born offers an opportunity to illustrate this cultural blindness. Considering what has been reported about childbirth in pre-literate and pre-agricultural societies challenges the assumption that midwifery is universal, and that seeking assistance when giving birth is a human characteristic.

Pre-midwifery societies

One of the most useful written documents about childbirth in pre-literate and pre-agricultural societies is the 2008 book by Daniel Everett, *Don't Sleep, There Are Snakes*.[1] One of the reasons why this document is valuable is that the author – a male missionary and a linguist – originally had no personal curiosity about childbirth. He reports what he learnt about life and language among the Pirahãs, who live in the Brazilian Amazonian jungle by the Maici River. Neither the blurb presenting the book nor the numerous published endorsements mention what Everett wrote about the way Pirahã women give birth. It is only in the middle of a chapter entitled 'Material Culture and the Absence of Rituals' that precious information is incidentally provided. The report is authoritative, since Daniel Everett spent nearly 30 years of his life among the Pirahãs.

The Pirahãs have kept many characteristics of pre-agricultural (Paleolithic) societies. Preparing and planting fields of manioc is a new development, introduced by the American linguist Steve Sheldon, who preceded Daniel Everett in the study of the local language. The use of imported machetes is also recent.

In such a context, Daniel Everett's account of childbirth deserves our attention. A Pirahã woman usually gives birth by herself, and there is no special place to be in labour. It depends upon the season.

In the dry season, when there are beaches along the Maici, the most common form of childbirth is for the woman to go alone, occasionally with a female relative, into the river up to her waist, then squat down and give birth, so that the baby is born directly into the river. This is cleaner and healthier, in their opinion, for the baby and the mother.

Not only is there no concept of a birth attendant, but labouring women's use of water is described, suggesting that it may be an instinctive human behaviour. This is probably the only written document about an ethnic group where birth under water is common.

During the time Steve Sheldon spent among the Pirahãs, a woman died when giving birth. It was a breech presentation. It is significant that she was alone giving birth at the river's edge. It is also significant that when the labouring woman's screams were heard, Steve Sheldon wanted to go and help her. He was told not to. For the local people, who probably understand that one cannot help an involuntary process, any interference would have increased the risk of a disaster. The reflex of the visitor, on the other hand, was to *do* something. This is an illustration of our powerful post-Neolithic cultural conditioning. What kind of help might an American linguist have provided in this particular case, in the middle of the Amazonian jungle?

There are obvious common points between the report by Daniel Everett and what Marjorie Shostak and her husband Melvin Konner wrote about the 'solitary and unaided births' among the African hunters-gatherers the !Kung San, who live in the Kalahari desert in Namibia, Botswana and Angola.

> *A woman feels the initial stages of labour and makes no comment, leaves the village quietly when birth seems imminent, walks a few hundred yards, finds an area in the shade, clears it, arranges a soft bed of leaves, and gives birth while squatting or lying on her side – on her own.*[2]

Having spent nearly two years with the !Kung in the early 1970s, Marjorie Shostak became intimate with some local women, particularly Nisa, who gave birth to four babies in the 1930s and 1940s, at a time when the local people were still living as their ancestors had done before them – gathering wild plant foods and hunting wild animals in their semi-arid environment.[3] This is an extract of a conversation with Nisa:

> *I have always refused to give birth with anyone there. I have always wanted to go alone. Because, although people try to help you by holding and touching your stomach, they make it hurt more. I did not want them to kill me with any more pain. That's why I always went by myself.*

In 1978, at an 'ethno-obstetrics' conference in Gottingen, Germany, I had the opportunity to watch films by Wulf Schiefenhövel (from the Max Planck Institute), which he made among the Eipos, who live in the highlands of New Guinea.[4] The Eipos were not perfectly representative of a pre-agricultural ethnic group, since they had gardens and pigs. However, they still had Paleolithic characteristics. Birth was not socialised and, in the films made discreetly by Wulf and his wife, women are seen giving birth in the bush, without any assistance.

One scene is highly significant. A labouring woman goes into the bush with her mother. They prepare a fern bed. Then the mother disappears, after giving her daughter a gentle pat, as if to say 'Good luck!' Within seconds the woman's mother has transmitted two eloquent, non-verbal messages. The first is: 'From now on I cannot help you'. The second is: 'My presence might make the birth more difficult'. Soon after we see the young woman giving birth through a typical 'fetus ejection reflex'. This means that there is no room for voluntary movements. A similar fetus ejection reflex may occur among modern women in exceptionally rare situations that are not, in general, culturally acceptable. It can occur, for example, when there is nobody around the labouring woman, apart from one experienced motherly and silent midwife keeping a low profile, just sitting in a corner and knitting. The documents presented by Wulf Schiefenhövel are unique and therefore precious because, in general, a film cannot show an authentic fetus ejection reflex: if a modern labouring woman knows that there is a camera around the fetus ejection reflex is transformed into a second stage of labour with a need for voluntary movements.

Where was the mother of the labouring woman after she had helped her daughter to prepare a fern bed in the bush? We can assume that she was not far away; invisible but aware of what was happening. She probably remained in a position to protect the space against the presence of a wandering human being or other animal. This might be the root of midwifery: a mother figure discreetly protecting the birthing place.

Biased knowledge

Our knowledge of life, including childbirth, in pre-agricultural and pre-literate societies, is limited and biased by the fact that only human groups living inland, well above sea level, have been studied. The Pirahãs live thousands of miles uphill from the mouth of the Amazon river in the Atlantic Ocean. The Kalahari, where the !Kung San people have lived for 20,000 years, is far from the sea, and the Eipos live in the highlands of New Guinea. In general, groups of hunter-gatherers that could be studied by anthropologists live inland. This is true of the Ache, in Paraguay, or the Hiwi, who live in the savannahs of western Venezuela and eastern Colombia. When sea levels rose at the end of the last Ice Age around 10,000 years ago, we lost the chance to observe hunter-gatherer populations living close to the oceans. This bias has been reinforced by the effects of the colonisation of the American continents by Europeans. We know more about the Sioux of the Dakotas, Minnesota and Iowa, or about the Pirahãs or the Hiwi, than about those who were living close to the oceans.

Such bias is important, since a great part of humanity had probably been living by the sea. It is partly by hugging the coasts that our ancestors have colonised the whole planet. Genetic studies demonstrating that some Amazonian Native Americans are related to indigenous Australians, New Guineans and Adaman Islanders are keys to understand the marine aspect of human beings.[5] In the current scientific context, we have realised that *Homo sapiens* has all the physiological characteristics of a primate adapted to the coast.

For example, having an enormous, highly-developed brain requires daily iodine.

It is significant that, according to ATA (American Thyroid Association), women should take a daily supplement (150 μg) of iodine before conception, and during pregnancy and lactation. According to a British study, a daily supplement of iodine when the mother was pregnant is associated with a 1.22 average increased IQ.[6]

Iodine is a 'brain selective nutrient' because of its essential role in thyroid hormone production, which in turn is needed for normal brain development. It is notable that iodine is the only nutrient for which governments legislate supplementation, so that iodination of table salt is mandatory. In spite of such widespread legislation and many public health strategies (such as dripping iodine into the water of Chinese irrigation ditches), iodine deficiency is the most common nutritional deficiency at a planetary level. It is the leading cause of preventable intellectual disabilities. It is difficult for humans to obtain sufficient iodine if their diet is lacking in seafood. These issues are particularly serious when considering the increased needs during pregnancy and lactation. Recently it was revealed that salts used in brine for pickling food – and also commercialised 'sea salt' – have no added iodine.

The long history of iodised salt is in itself highly significant. French chemist Jean-Baptiste Boussingault recommended iodised salt for goitre prophylaxis in the 1830s. He had demonstrated that the 'marine smelling' fluid from the salt deposits in the goitre-free Antioquia (South America) could reduce endemic goitre in the neighbouring regions.

There are many other human characteristics confirming our adaptation to the coast. Another one relates to the specific nutritional needs of the brain, particularly the developing brain. In simple terms, the human body is not very effective at making a molecule of fatty acid that is essential for the brain. This molecule is pre-formed and abundant in seafood, but is not found elsewhere. Let us translate that into the language of biochemists: one of the most significant characteristics of *Homo sapiens* is the association of an enormous, highly developed brain with a weak delta 4 enzymatic system of desaturation.[7] This suggests that *Homo sapiens* is adapted to an environment providing pre-formed docosahexaenoic acid (DHA).

The genetic perspective is also challenging when we are looking for evidence that *Homo sapiens* is physiologically adapted to the seashore, since only inland populations are easily studied. We understand how early inland migrants adapted to other environments, including life at altitude.[8] For example, we understand how the Tibetan version of the gene called EPAS1 spread rapidly about 30,000 years ago as an adaptation to the lack of oxygen. We know about the genome – and even the epigenome[9] – of archaic humans (including Denisovans) who were living about 40,000 years ago in the remote Denisova Cave in the Altai Mountains in Siberia. But geneticists and epigeneticists cannot provide precise data about those living along the Asian coasts at that time.

In spite of these limitations and biases to our knowledge of pre-agricultural *Homo*, we have sufficient information to claim that the participation of a birth attendant is not a universal human characteristic.

The specifically human handicap

We also have good reasons to challenge the assumption that mechanical factors are the main reasons for difficult births in our species. There are women with no morphological particularities who give birth quickly without any difficulty. There are anecdotes of women who give birth before realizing that they are in labour. There are countless anecdotes of teenagers who, at the end of a hidden or undiagnosed pregnancy, just go to the toilet and give birth within minutes.

These facts alone suggest that the main reasons for difficult human births are not related to the shape of the body.

The best way to clarify the nature of the specifically human handicap during the period surrounding birth is to consider the case of civilised modern women who have given birth through an authentic fetus ejection reflex. It is exceptionally rare in the context of socialised birth. The birth is preceded by a very short series of irresistible, powerful and highly effective uterine contractions, without any room for voluntary movement. After an authentic fetus ejection reflex, many babies are 'born with the caul', which means with intact membranes.

The important point is that when the 'fetus ejection reflex' is initiated, women are obviously losing neo-cortical control – control by the thinking brain, the brain of the intellect, the part of the brain that is highly developed only among humans. Women may suddenly talk nonsense. They can behave in a way that usually would be considered unacceptable for a civilised woman, for example screaming, swearing, biting the midwife… They can find themselves in the most unexpected, bizarre, often mammalian, quadrupedal postures. They seem to be 'on another planet'. A reduced neocortical control is obviously a prerequisite for an easy birth among humans. In other words, the main reason for the human handicap during such a physiological process is the inhibitory effect of an active, powerful neocortex.

It is easy to explain why the concept of fetus ejection reflex is not understood after thousands of years of socialisation of childbirth. It is precisely when birth seems to be imminent that the birth attendant has a tendency to become even more intrusive. If, for example, the labouring woman says: 'Am I going to die?' 'Kill me!' or 'Let me die…', instead of keeping a low profile, the well-intentioned birth attendant usually interferes, at least with reassuring, rational words. These rational words can interrupt progress towards the fetus ejection reflex. The reflex does not occur if there is a birth attendant who behaves like a coach, or an observer, or a helper, or a guide, or a 'support person'. And it is exceptionally rare when the baby's father participates in the birth. The fetus ejection reflex can also be inhibited by eye-to-eye contact or the imposition of a change of environment, as would happen, in our modern world, when a woman is transferred to a delivery room. It is inhibited when the intellect of the labouring woman is stimulated by any rational language, for example if the birth attendant says: 'Now you are at complete dilation. It's time to push.' During an authentic reflex, the mother is in a quasi-ecstatic state and does not even realise that the baby is coming. Any interference tends to bring her back 'down to Earth', and tends to transform the fetus ejection reflex into a second stage of labour, which involves voluntary movements.[7]

The term 'fetus ejection reflex' was coined by Niles Newton in the 1960s, when she was studying the environmental factors that can disturb the birth process in mice.[10] She had revealed the importance of cortical activity, even among non-human mammals. Twenty years later, with her support,[11] I suggested that we save this concept from oblivion; I was convinced it could be a key to facilitating a radically new understanding of the specifically human difficulties during the process of parturition.[12]

The deep-rooted belief that the presence of a specialised person is the basic need of labouring women and newborn babies is the first obstacle to overcome. In our societies unassisted deliveries occur occasionally by accident. Either the midwife could not arrive in time, or the mother could not reach the hospital in time. Because such births are easy, one might anticipate that it is an opportunity for some women to reach the climax associated with a fetus ejection reflex. It is not so in most cases, because of the cultural conditioning that a woman is unable to give birth by herself. For example, if the husband/partner is around, he is usually in a state of panic wondering what he must do and who will 'deliver the baby' and who will cut the cord.

We should keep in mind that many human abilities are obscured by neocortical activity. The main obstacle to overcome must be understood in the enlarged framework of a deep-rooted lack of interest in this essential aspect of human nature. For example, the capacity to adapt to immersion and have coordinated swimming movements when submerged disappears around the age of three or four months, when the neocortex is reaching a certain degree of power.[13] An ingenious experiment has demonstrated that human olfactory abilities are obscured by neo-cortical activity. The principle of this experiment was simply to explore the sense of smell after neo-cortical disinhibition by alcohol consumption.[14]

In realising that midwifery is not a universal human characteristic and that neocortical activity is an obstacle to easy birth in our species, we have taken significant steps in our understanding of human nature.

CHAPTER 3

A useless question

After analysing the reasons why our question may be considered ludicrous, and after analysing the reasons why it might be considered sensible, we'll look at the point of view of those who would claim that this question is useless and anachronistic anyway, since caesarean birth is the future. It is still common to talk about natural selection – or survival of the fittest – where human beings are concerned. However, it is an undisputable fact that modern medicine, particularly obstetrics, has neutralised the laws of natural selection. Until recently, only women who had a tendency to give birth easily had many children without risking their life. Today the number of children per woman depends on factors other than her capacity to give birth. These facts suggest an irreversible tendency towards an increased need for caesareans.

Towards a new understanding of normality

As the concept of 'natural selection' may appear purely theoretical, we must use other examples to persuade readers that the caesarean will become the 'normal' mode of birth in a predictable future. The simplest and fastest way is to analyse graphs demonstrating the continuous tendency towards more caesarean sections. In developed countries, the caesarean rate was in the region of 2 per cent in the middle of the twentieth century. It is now nearer to 33 per cent in countries such as the United States, Germany, Australia and New Zealand. Caesarean rates have increased even more quickly in emerging countries such as China and most Latin American countries. Using these data as a starting point, all projections and extrapolations suggest that, at a planetary level, the abdominal route will be the 'normal' way to be born during the second half of the twenty-first century.

Technical advances have made caesareans faster, simpler and less risky. When I did my first caesarean sections in the 1950s, the operating time was in the region of an hour, and we usually needed one or two units of blood to compensate for the blood loss. Today, thanks in particular to the technical simplifications introduced by Michael Stark in the 1990s, the usual operating time is in the region of 20 minutes and blood loss is dramatically reduced. General anaesthesia is no longer necessary in most cases.

It is undisputable that, in modern well-equipped and well-organised hospitals, the caesarean has reached a degree of safety quite impressive to practitioners of my generation. This is illustrated by eloquent statistics. In a report of all women in Canada (excluding Quebec and Manitoba) who gave birth between April 1991 and March 2005, the planned caesarean group for breech presentation comprised 46,766 women.[1] No mothers died in this group, while 41 of the 2,292,420 women died in the planned vaginal delivery group (maternal mortality rate 1.8 per 100,000 births: the differences are not statistically significant). Such data are useful since the risks of maternal death related to caesarean births are difficult to evaluate. These caesareans were not related to maternal health problems, but to the breech presentation. In most statistics, the risks appear three to four times higher after a caesarean than after a birth by the vaginal route. But the studies are hampered by the fact that a number of women having caesareans have conditions, pregnancy complications, and delivery complications that are themselves associated with increased maternal mortality.

There are even published evaluations of the effects on birth statistics of anticipated increased rates of caesareans. A study took into account that, among the four million births a year in the United States, about three million occur at 39 weeks or after. It was estimated that if all American women reaching 39 weeks were giving birth by caesarean, either before labour starts or when labour is established, 9,462 cases of neonatal encephalopathy a year would be prevented.[2] Statistics from the largest obstetrical centre in Shanghai, China, lead to similar conclusions. Among the 66,226 women who gave birth to their first babies at term in that centre between 2007 and 2013, a quarter had chosen a scheduled caesarean (only flexed head presentations were included in the series). The risks of neonatal encephalopathy were 4.5 times lower after caesarean delivery on maternal request than when a vaginal birth had been originally planned.[3]

Other reasons to increase the rates of caesareans are suggested by the results of long-term epidemiological studies. A typical example is offered by a Chinese study published in 2010. The objective was to evaluate the likelihood of psychopathological problems in childhood in relation to how babies are born.[4] This huge study (more than 4,000 first babies born at term) was conducted in south-east China, among a population where the overall rate of caesarean sections was 56 per cent. Three groups of children were assessed. Children in the first group were born by the vaginal route without forceps (or ventouse). Those in the second group were born with forceps (or ventouse). Those in the third group were born by programmed pre-labour c-section on maternal request. Children in the third group, born by pre-labour c-section, had the lowest risk of psychopathological problems. After controlling for many associated factors, the differences remained highly significant. One of my interpretations is that, in the context of modern China, almost all children had probably been exposed to synthetic oxytocin, except those born by pre-labour caesarean section.

This interpretation is not inspired by data included in the published article. As is usual in medical literature, synthetic oxytocin is not mentioned. It is a paradoxical situation, since synthetic oxytocin is widely used in developed countries and emerging countries as well. It is the most common medical intervention in childbirth at a planetary level. Until now there has been a lack of research about the side effects of this component of pharmacological assistance during the different phases of labour, although a great variety of such effects is plausible. A new generation of research might provide new reasons to prefer in-labour caesarean sections to long hours of pharmacological assistance. This might have significant impact on modes of birth, particularly in countries – like France – where the caesarean rate has remained stable for several years after a spectacular surge. These are countries where pharmacological assistance is widely used, without taking into account its probable long-term side effects.

For example, there are theoretical reasons to test the hypothesis that the increasing incidence of breastfeeding difficulties and the frequent earlier than desired cessation of breastfeeding are related to the use of synthetic oxytocin during labour.

Some epidemiological studies support this hypothesis.[5,6] There have been video explorations of the effects of synthetic oxytocin on primitive neonatal reflexes, and therefore the initiation of breastfeeding.[7,8] The main limitation of these scarce preliminary studies is that, in the context of developed countries, it is difficult to dissociate possible side effects of synthetic oxytocin from the effects of other components of pharmacological assistance during labour, particularly the effects of epidural anaesthesia. This is why studies should be undertaken in emerging countries, where synthetic oxytocin is widely used but there are no epidural services.[9]

Finally, there are inescapable questions to be faced about the probable evolution of the human capacity to give birth, even without referring specifically to the laws of natural selection. Today, with the help of epigenetics, it is understood that when a physiological function is underused it has a tendency to become weaker and weaker from generation to generation. In an age of pharmacological assistance during labour and simplified techniques of caesarean section, the oxytocin system is an example of a physiological function that has suddenly become underused: until recently, it was impossible to give birth to babies and placentas without an intense activation of this system. From another perspective, practitioners who routinely ask how a pregnant woman was herself born have come to the conclusion that there is a correlation between the way a woman was born and how she gives birth to her own children. Of course it is only a correlation, and many factors are involved. However, there are genuine reasons for raising questions about the human capacity to give birth after further decades of industrialised obstetrics.[10]

Let us dare to go one step further and suggest this human ability might already be weakening. A recent accumulation of data suggests that factors other than an inappropriate hospital environment must be taken into account to explain the increased difficulties of modern births, since these increased difficulties are particularly significant when considering the case of home births. The conclusions of an American study that looked at 'Changes in Labor Patterns Over 50 Years' are highly significant.[11] The authors compared a first group of nearly 40,000 births that occurred between 1959 and 1966 and a second group of nearly 100,000 births that occurred between 2002 and 2008. They only looked at births of one baby at term, with head presentation and spontaneous initiation of labour. After taking into account many factors such as the age, height and weight of the mother, it appeared that the duration of the first stage of labour was significantly longer in the second group. It was two and a half hours longer in the case of a first baby, and two hours in the other cases. For practitioners of my generation this study demonstrates the obvious. It justifies the questions we are raising.

Preliminary signs of a new normality

The tendency to consider the caesarean as the 'normal' way to give birth is already detectable. It is significant that a great number of female obstetricians with an uncomplicated pregnancy at term choose a scheduled caesarean delivery for the birth of their own babies.[12,13] The concept of caesareans on demand is a significant step in our history. It is accepted, and even supported, by some of the most influential medical organisations. It is also significant that, as early as 2003, the American College of Obstetricians and Gynecologists' ethics committee published a statement judging elective caesareans ethical. W. Benson Harer Jr, medical director of the Riverside

County Regional Medical Center in Moreno Valley, California, commented on this statement: 'I think it is a step to where we're going. And my guess is that as increasing evidence comes out, it will probably become a more accepted procedure'.[14] At the same time, in the UK, the National Institute for Clinical Excellence (NICE) issued preliminary guidelines making clear that doctors should not refuse a woman the right to have a caesarean, but that the reasons for the request should be sought, recorded and discussed.

Pre-labour versus in-labour caesareans

When the caesarean section is accepted as a 'normal' way to be born, the focus is different. The main questions are about the timing of the operation. When is the best time to perform the operation? From that perspective, there are three kinds of caesareans: pre-labour c-sections, in-labour non-emergency c-sections, and in-labour emergency c-sections. We must reconsider the usual classifications, which confuse the terms 'pre-labour' and 'elective', and also 'in-labour' and 'emergency': it is possible to plan an in-labour c-section, and an in-labour operation can be decided on and performed before the birth becomes an emergency.

In the current scientific context, we should also go beyond the usual classifications according to the route of birth. From the point of view of the baby, it appears that the main differences are between pre-labour caesarean births and others (whatever the route).

It is easy to anticipate differences between babies born pre-labour and those born by the other routes. In particular we know that there are increased risks of respiratory difficulties after pre-labour birth, which makes sense since we now understand that the fetal lungs themselves provide a signal to initiate labour: surfactant, a protein within the lungs, serves as a labour hormone that indicates to the mother's uterus that the fetal lungs are mature enough to withstand the critical transition from life in fluid to air breathing.[15] Furthermore, the well-known role of maternal and fetal stress hormones in the maturation of the baby's lungs is relevant. Everybody knows that when a premature birth is considered imminent the mother is given analogues of stress hormones (corticosteroids) to mature the baby's lungs. The stress of labour is associated with the release of endorphins, which induce the release of prolactin. One of the many effects of prolactin is to participate in lung maturation.[16] We must also take into consideration the effects of the fetal stress hormone noradrenaline, which is released during labour. Fetal noradrenaline has multiple roles to play, including protection against lack of oxygen during uterine contractions and lung maturation.

This is an important point: the multiple negative effects of stress deprivation among babies born by pre-labour c-section have been underestimated until recently. For example, it has been demonstrated that, under the effect of noradrenaline, the sense of smell has reached a high degree of maturity at birth among babies born by in-labour c-section. The principle of a Swedish experiment was to expose babies to an odour for 30 minutes shortly after birth and then to test them for their response to this odour (and also to another odour) at the age of three or four days.[17] Since the concentrations of noradrenaline had been evaluated, it was possible to conclude that fetal noradrenaline released during labour is involved in the maturation of the sense of smell. We must emphasise the paramount role of the sense of smell immediately after birth. I mentioned in the 1970s that the sense of smell is the main guide towards the nipple during the hour following birth.[18,19] It has been demonstrated that it is mostly through the sense of smell that a newborn baby can identify his mother (and, to a certain extent, that the mother can identify her baby).

There has recently been an accumulation of data to support the practice of waiting, whenever possible, for the onset of labour before performing a caesarean. Many unexpected differences between pre-labour and in-labour caesareans have been demonstrated in human studies. One such difference has been found in the evaluation of adiponectin concentration in cord blood of healthy babies born at term (adiponectin is a metabolic hormone involved in fat metabolism). The concentration is significantly lower after pre-labour caesarean compared with in-labour caesarean or vaginal birth.[20] These data suggest a mechanism in which stress deprivation at birth might be a risk factor for obesity in childhood and adulthood. Other important data concerns the milk microbiome. There are significant differences between the milk of mothers who gave birth by pre-labour caesarean and those who gave birth by in-labour caesarean or vaginally.[21] These results suggest that there are other factors than the operation itself that can alter the process of microbial transmission to milk. Similar differences were found by a Canadian study of the gut flora of four-month-old babies.[22] Joanna Holbrook and her team in Singapore suggest interpretations for these surprising data. They collected fecal samples from 75 babies at the age of three days, three weeks, three months and six months, and they evaluated the degree of adiposity at 18 months. It appears that, apart from the route of birth and exposure to antibiotics, a shortened duration of pregnancy tends to delay the maturation of the gut flora: one week more or less in the duration of pregnancy is associated with highly significant differences. A pre-labour caesarean implies the association of all the known factors that can delay the maturation of the gut flora. This study is all the more important as it also reveals that a delayed maturation of the gut flora is a risk factor for increased adiposity at the age of 18 months.[23]

Human studies have also included evaluations of the concentrations of melatonin in the cord blood. It is low after pre-labour births.[24] This is an important point, since melatonin (the 'darkness hormone') has protective anti-oxidative properties. Furthermore, it confirms that the 'darkness hormone' is involved in the birth process. There are other reasons why the role of melatonin during labour is topical, at a time when we are learning about a synergy between its receptors in the uterus and oxytocin receptors.

In general, a baby born after a pre-labour caesarean is physiologically different from the others. For example, babies born pre-labour tend to have a lower body temperature than the others during the first 90 minutes following birth.[25]

In spite of possible interspecies differences, we must seriously consider animal experiments suggesting that the stress of labour influences brain development. Studies have demonstrated that the birth process in mice triggers the expression of a protein (uncoupled protein 2) that is important for hippocampus development.[26] In humans, the hippocampus is a major component of the limbic system. It has been compared to the 'conductor of an orchestra' directing brain activity. It has also been presented as a kind of physiological GPS system, helping us navigate while also storing memories in space and time: the 2014 Nobel Prize for physiology and medicine was awarded to three scientists who have studied this important function of the hippocampus. Studies in rats have suggested that oxytocin-induced uterine contractions reverse the effects of the important neurotransmitter GABA: this primary excitatory neurotransmitter becomes inhibitory.[27] If uterine contractions affect the neurotransmitter systems of rats during an important phase of brain development, why would the same not occur in humans?

This overview of the multiple effects of stress deprivation in the case of pre-labour birth suggests that the ideal caesarean should be performed during labour, before the stage of a real emergency. Other reasons to avoid pre-labour caesareans will probably appear in the near future. It seems that the prevalence of placenta praevia is significantly increased only in the case of a pregnancy following a pre-labour caesarean.[28] There is already an accumulation of data confirming the negative effect of pre-labour caesarean on breastfeeding prevalence, particularly at the phase of initiation of lactation.[29,30] We must also keep in mind that emergency caesareans are associated with comparatively bad short-term statistics. Furthermore, they are associated with negative long-term outcomes. For example, according to an American study, women with a full-term second-stage caesarean have a hugely increased rate of subsequent premature births (13.5 per cent) compared to those who had a first-stage caesarean (2.3 per cent) and to the overall national rate (7–8 per cent).[31]

When the new normality becomes established, we'll enter a radically new phase in the history of medicine. New questions will preoccupy us. It will probably be common to wonder how we can compensate for the microbial deprivation of a baby born via the abdominal route into the sterile environment of an operating room.

What if...?

When we use rational language to explain that caesarean section might become the 'normal' way to be born, many react with the immediate gut feeling that this scenario is not acceptable. Until now this has been a mostly subjective reaction. But what if, in the near future, that gut feeling is backed up by emerging and fast-developing scientific disciplines? These include the branch of epidemiology we call 'primal health research', metagenomic bacteriology, epigenetics, and several fast-developing branches of physiology and evolutionary biology. They have the power to broaden our horizons and oblige us to think in terms of the future of our species. If we also take into account that modern physiology offers a new understanding of the basic needs of labouring women, we have serious reasons to look at other scenarios than the one considered logical until now.

The main obstacle to a new awareness is specialisation. This is why we need an interdisciplinary approach. However, in order to summarise what emerging and fast-developing disciplines can offer, we need a point of departure. This point of departure should be 'primal health research', the branch of epidemiology that explores correlations between what happens during the 'primal period' and what happens later on in terms of health and personality traits: it provides hard data collected in the Primal Health Research Database. The other emerging disciplines inspire interpretations and theories. A frequent mistake is to 'put the cart before the horse' and start with theories. The studies of risk factors for autism are an example. Everybody, all over the world, has heard of theories suggesting that certain infant vaccinations are risk factors for autism. The theories have never been backed up by epidemiology. At the same time a dozen scientifically valuable epidemiological studies have shown that the period of birth appears to be critical for gene/environment interaction in the case of autism. There has been a tendency to ignore these data. Are theories more attractive than facts?

CHAPTER 4

primalhealthresearch.com versus NIH

At the end of 2014, nearly 30 years after the publication of *Primal Health* and the beginning of the Primal Health Research Database, the US Congress called for a review of the way the budget of the National Intitutes of Health (NIH) is used. The panel included authoritative scientists from the Institute of Medicine and the National Research Council. The group agreed that there was a need to study how exposures early in life affect human health.

The reason for the review was an ambitious plan hatched in around the year 2000 to follow the health of 100,000 US children from before birth to age 21. After more than $1.2 billion had been spent on the study it became clear that it was not feasible and the NIH Director announced that the study would be dismantled.

We must look at the possible positive effects of this story. Any spectacular event – even scandalous – can trigger new awareness.

From knowledge to awareness

The emergence of a new awareness is mysterious. How can we reduce the distance between knowledge and awareness? Reaching a new awareness may be a sudden and fast process. The French term 'prise de conscience' is appropriate because it makes people think of 'prise de courant' (turning your electricity on at the electric outlet). Triggering factors are unpredictable. However, in the context of the twenty-first century, one cannot imagine triggering factors independent of spectacular scientific advances. We must rely on scientific knowledge to induce necessary transformations of Homo sapiens. Homo sapiens has not been programmed to think long term and in terms of survival of the human species. Before the advent of agriculture and animal husbandry, our ancestors could live from day to day. Then they had to think in terms of seasons. Today we should be thinking in terms of decades, centuries and even millennia.

The efficacy of modern medicine is an example. From an individual and short-term perspective, modern medicine can be considered miraculous, even compared with the 'medical art' I was taught as a medical student in around 1950. However, our contemporaries do not dare to consider the crucial fact that medical practices counteract natural selection. If we were realistic, we would acknowledge that if we keep going in the same direction, a time will come when the life of nearly all human beings is dependent on medical institutions. How can we reduce the distance between this dismal scenario and an awareness of its inevitable effects in terms of the survival of our species? The urgent first step is to develop the human capacity to think long term.

A useful tool

It is in this context that we present the Primal Health Research Database as a tool to train ourselves to think long term. It is imperative, particularly among those involved in childbirth. There are deep-rooted and understandable reasons why most midwives and obstetricians cannot easily see beyond the perinatal period. Traditionally everybody was happy when mother and baby were alive and healthy after what is often considered the most dangerous phase of human life. There has not been any significant paradigm shift until now. The usual modern criteria for evaluating the practices of obstetrics and midwifery are still short term. In medical language, they are called 'perinatal and maternal mortality and morbidity rates'. During their endless discussions about the comparative safety of hospital births and home births, both medical circles and 'natural childbirth' groups focus on these conventional criteria.

Our database includes published studies that explore correlations between what happens during the 'primal period' and what happens later on in terms of health and personality traits. The primal period includes fetal life, the period surrounding birth and the year following birth. The first time I heard of attempts to detect risk factors for pathological conditions in the period surrounding birth was in July 1982 when I met Nikolaas Tinbergen. Interestingly, he had been awarded the Nobel Prize in 1973 as one of the founders of ethology: he was not an epidemiologist. We met in Oxford, at a conference organised by the McCarrison Society, after my presentation about 'Childbirth and the Diseases of Civilization'.[1] My objective, at that time, was to give theoretical reasons for anticipating a new generation of research. Nikolaas Tinbergen told me that, using his method as a 'field ethologist', he had detected risk factors for autism in the perinatal period. According to him, labour induction, difficult forceps delivery and anaesthesia during labour were risk factors (in a population of children born in the 1970s).

It is worth emphasising that, in our database, the oldest epidemiological study is dated 1984. It is a Norwegian study of the IQ of 18-year-old conscripts born by forceps. Today, countless other studies can be detected by selecting appropriate keywords. Some keywords make prospective research possible. This is the case, for example, of keywords such as 'birth complications', 'forceps delivery', 'ventouse', 'labour induction', 'caesarean' and 'synthetic oxytocin'. Other keywords make retrospective research possible. This is the case, for example, for 'autism', 'allergic diseases', 'allergic rhinitis', 'hay fever', 'asthma', 'anorexia nervosa', 'drug addiction', 'diabetes type 1', 'obesity', 'suicide', 'IQ' and 'criminality'. An overview of all these studies leads to the conclusion that this emerging branch of epidemiology has been developing at a high speed since the 1980s. The other conclusion is that the long-term consequences of modes of birth have been ignored until recently … except by a small number of intuitive pioneers.

It is only through an overview of the whole database that one can reach such conclusions. When focusing on one particular study, there are always reasons to discuss its scientific value and its conclusions. It is relevant to keep in mind that 'correlations' and 'risk factors' brought to the fore by epidemiologists do not always imply that there is a cause and effect relationship, even if statisticians are more effective than ever at taking into account associated and confounding factors. Furthermore, when researchers in different parts of the world and using different protocols provide concordant results, the cause and effect relationship is highly probable. Ideally, we should rely on longitudinal studies following up populations. But we must start with the kind of research that is feasible with limited budgets. The cost of primalhealthresearch.com is the time of one volunteer exploring the scientific and medical literature.

Concordant results

The keyword 'autism' offers a good example of concordant results provided by a great number of studies. Autism is not a purely genetic disease: it is, therefore, useful to identify critical periods for gene/environment interaction and, if possible, the nature of the responsible environmental factors.

We have already mentioned the perspective of Nikolaas Tinbergen who – as an ethologist – was studying autistic children in their family environment in around 1980. In 1991, Ryoko Hattori, from Kumamoto, Japan, published in an authoritative medical journal a thought-provoking study of the risks of autism according to place of birth. She revealed that children born in a hospital where the 'Kitasato University Method' was routine were at increased risk of becoming autistic.[2] Although it does not appear in the title of the article, the main characteristic of this method was labour induction a week before the due date. After being alerted by Nikolaas Tinbergen, I found the study so important that I went to Kumamoto in the 1990s and met Ryoko Hattori.

After encountering these pioneering works, I innocently expected a flood of epidemiological studies on the same topic in the near future. I eventually expressed my impatience in 2000 by introducing in an authoritative medical journal the concept of 'cul-de-sac epidemiology'.[3] Taking the example of autism (and drug addiction), I referred to studies we prefer not to look at, not to enlarge, not to replicate, and not to quote after publication. I'll never know if my paper had a triggering effect, but a series of valuable studies were published between 2002 and 2006 confirming, in general, the preliminary conclusions. In 2002, a study was published involving the whole Swedish population born over a period of twenty years (more than two million births!). According to this enormous study a caesarean birth and a low Apgar score were among the risk factors for autism.[4] The same year a Canadian study was published comparing 78 children with autistic spectrum disorder and 88 unaffected siblings: children with autism had higher rates of birth complications.[5] In 2004, an Australian study compared 465 autistic subjects born in Western Australia over a period of 15 years with their 481 siblings and 1,313 controls. Labour induction, caesarean section (particularly pre-labour c-section), low Apgar score, fetal distress during labour and birth complications in general appeared as risk factors. It is notable that being first born also appeared to be a risk factor: the first delivery is often more difficult than subsequent births.[6] A Danish study published in 2005 associated 698 autistic subjects with 25 controls for each: low Apgar score and breech presentation appeared as risk factors.[7] A study from Israel published in 2006 interviewed hundreds of mothers about prenatal, perinatal and neonatal complications; the interesting point is that there were no prenatal differences between the groups, but more birth complications in the group of mothers of autistic children.[8]

The list of valuable studies on autism from a primal health research perspective has recently expanded. Whatever the country and the research protocol, none of these studies has contradicted the previous findings. This is the case in a study from Zhengzhou, China, which looked at pregnancy complications, birth asphyxia, and premature rupture of the membranes.[9] It is also true of a twin study from California.[10] The new research is dominated by a huge study from North Carolina, which combined data from about 625,042 births and school records, including the cases of more than 5,500 children diagnosed as autistic.[11] Compared to children born to mothers who received neither labour induction nor augmentation, children born to mothers who were induced and augmented, only induced, or only augmented were at increased risk for autism after controlling for potential confounders related to socioeconomic status, maternal health, pregnancy-related events and conditions, and birth year. The observed associations between labour induction/ augmentation were particularly pronounced in male children. A study involving the whole Danish population between 2000 and 2008 has confirmed that labour acceleration is a risk factor among boys.[12] It is notable that studies looking at maternal obesity and maternal diabetes as possible risk factors for autism did not take into account labour induction and labor augmentation, although the need for pharmacological assistance is undoubtedly influenced by the metabolic type the mother belongs to.[13,14]

Genesis of non-communicable diseases

This overview of entries reached through the keyword 'autism' provides an opportunity to mention an essential function of our database. Until now, when considering the origin of pathological conditions not purely genetic, it has been usual to raise only two types of question. The first type leads to endless discussion of the comparative roles of genetic and environmental factors: the modern way to continue the classical 'nature versus nurture' debate. The second type leads to trying to identify the genes involved in particular diseases. From a practical perspective, both groups of questions are of limited interest. It is more useful to provide data in terms of a critical period for gene/environment interaction. Our database has become a unique tool for providing clues about the critical periods for the genesis of diseases and personality traits. From an exploration of the whole database, it appears that often the *nature* of an environmental factor is less important than the *period of exposure* to this factor.

The importance of the concept of timing is obvious when considering the example of autism. A common point in the studies previously discussed is the detection of risk factors at birth. But the nature of the perinatal factors is vague and dependent on the variables researchers had at their disposal. For example, the huge Swedish study could not use the variable 'labour induction', which was not included in the national birth registry before 1991. But in studies that could take into account 'labour induction', it always appeared as a risk factor.

Some studies, particularly the Australian one, provide useful negative findings, suggesting that what happens before the birth has limited effects. Subjects who will become autistic have the same average birth weight, the same average head circumference at birth, and the same placental average weight as the others. In one study pre-eclampsia appears as a risk factor.[15] It is notable that the expression of the disease takes place in the perinatal period. Pre-eclampsia cannot be an important risk factor since, in many developed countries, its prevalence has decreased while the prevalence of autism has reached epidemic proportions. Two studies of the risks of autism spectrum disorders among children conceived *in vitro* (via IVF) suggest that the mode of conception does not significantly influence the risks of autism.[16,17] However, one study shows that it is possible that the technique of intracytoplasmic sperm injection (ICSI) is associated with a slightly increased risk. This technique, in contrast to conventional IVF, bypasses natural barriers to fertilisation, thereby increasing the possibility of the transmission of genetic defects. A study involving hundreds of California-based autistic children and hundreds of controls showed that mothers of autistic children were less likely to have taken iron supplements before, during and after pregnancy, and had a lower average daily iron intake.[18] Such results are surprising, because there are sound theoretical reasons why excess dietary iron might be a risk factor for autism.[19] The widespread prescription of iron supplements in pregnancy started in the 1980s, and since then the prevalence of autism has significantly increased. It is therefore unlikely that a low iron intake has played a significant role in the advent of the 'autism epidemic'.

Furthermore, what happens after the birth does not significantly influence the risks of autism. The mode of infant feeding does not seem to modify the risks either, and, in spite of much media speculation, we can make the same observation about infant vaccination. There is not a single valuable epidemiological study detecting correlations between MMR (Measles, Mumps, Rubella) vaccination and the diagnosis of autism, or detecting correlations with a vaccine containing a mercurial derivative.[20, 21, 22, 23, 24]

At a time when a spectacular increase in non-communicable diseases needs interpretation, it is worth emphasising – and it is not surprising – that risk factors are detected in the phase of modern lifestyle that has been the most dramatically transformed in recent decades.

Enlarging our horizon

Enlarging our horizon is not just about looking far into the future. It also means thinking in terms of civilisation, a specifically human dimension: our database also allows us to become familiar with this collective dimension, as epidemiologists often need huge numbers to detect statistically significant effects of early experiences.

The need to introduce the collective dimension and to think in terms of civilisation is clear when considering the birth process of non-human mammals. Among other mammals, when the birth process has been disturbed, the effects are dramatic and easily detected immediately at an individual level: the mother is not interested in her baby. This is the case of ewes giving birth with an epidural anaesthesia[25] or monkeys giving birth by caesarean.[26] Babies can survive only if human beings take care of them. Millions of women, on the other hand, take care of their newborn babies in spite of powerful interferences.

We can easily understand why it is much more complex in our species. Because human beings speak and create cultural milieus, there are situations when human behaviour is less directly under the effects of the hormonal balance and more directly under the effects of the cultural milieu. This is true of pregnancy and childbirth. When a woman is pregnant, she can express through language that she is expecting a baby, and she can anticipate her own maternal behaviour. Other mammals cannot do that. They have to wait until the day when they release a cocktail of love hormones to be interested in their babies. We should not conclude that we have nothing to learn from other mammals. They teach us which questions we should raise where human beings are concerned: in asking these questions we should always introduce the collective dimension via words such as 'civilisation' and even 'humanity'. This is why today the main questions are about the future of a humanity born by caesarean, or with epidural anaesthesia, or with drips of synthetic oxytocin…

One of the lessons of the primal health research perspective is that we should interpret anecdotes with extreme caution. For example, those who explore epidemiological studies should not be worried about one particular baby who was rescued by caesarean. The cultural milieu can compensate for many deprivations. Questions must be framed differently when human beings are concerned.

Limits to primal health research

All scientific disciplines have their limits. This is why, in the age of ultra-specialisation, interdisciplinarity is vital.

The limits inherent in the primal health research perspective are easily explained. It is well accepted that, in general, the 'golden method' of evaluating the ratio of benefits to risks of any human activity, particularly medical treatments and medical strategies, is the 'randomised controlled trial'. It means that researchers plan to study a population (for example a population with a specific disease). The first step is to divide this population into two (or more) groups by drawing lots ('at random'). A treatment (or a strategy) is used in one group, while another treatment is used in the other group. Then a follow-up makes possible a comparison between the two groups. For obvious reasons, this method cannot be used among humans to evaluate the long-term consequences of how babies are born. For example, one cannot ask a group of a thousand women to give birth by pre-labour caesarean and another group to give birth vaginally with a drip of synthetic oxytocin. Women would not agree to participate in such a study and ethical committees would veto the project. This is why our database cannot include this kind of research.

Animal experiments may offer opportunities to go one step further, although they have their limits: the findings in rats cannot necessarily be extrapolated to humans. Rats can give birth several times before reaching the age of one year. It does not take long to reach the equivalent of a dozen of generations among humans. There have already been studies of transgenerational effects of intervention during fetal life among rats, such as the effects of undernutrition, of exposure to corticosteroids or exposure to cocaine. The rare studies of the effects of caesareans looked only at the behaviour of the first generation.[27, 28]

There are current limits to the primal health research perspective that may change in the future. Until now, where childbirth is concerned, epidemiologists have contrasted vaginal birth and caesarean section. We now understand how important it is to contrast pre-labour and in-labour caesareans. The main differences, from the baby's point of view, might even be between pre-labour birth and birth with labour, whatever the route. A huge Scottish study of health in childhood in relation to the modes of birth is symbolic of the new generation of research we are expecting, although the vocabulary still belongs to the conventional paradigm, since 'planned caesarean delivery' appears as synonymous with pre-labour caesarean. According to this study, children born by caesarean before the beginning of labour are at increased risks of developing diabetes type 1, compared with all the other children. Differences are statistically highly significant, even after taking into account possible confounding factors, including maternal type 1 diabetes.[29] This data about an autoimmune disorder indicates the need for further studies of the risks of dysregulations of the immune system in relation to 'birth without labour'. There is a need, in particular, for a new generation of studies focusing on the risk factors for atopic diseases such as allergic dermatitis, allergic rhinitis and eczema.

Another limitation of primal health research will disappear with time. This is research about health conditions specific to old age, particularly neurodegenerative disorders such as Alzheimer's disease. Since this disease is usually diagnosed in people over 65 years of age, it is still difficult to explore risk factors related to modern obstetrics. There would be reasons, for example, to look at pre-labour caesareans if it is confirmed that the stress associated with uterine contractions is necessary for the expression of a protein that plays an important role in the development of the hippocampus.[30] It is well known that the hippocampus is one of the first regions of the brain to suffer damage in Alzheimer's disease.

CHAPTER 5

The driving force

Curiosity is the driving force behind scientific research. At any age, human beings are endowed with a huge innate capacity for curiosity. As an emotional state, curiosity may be induced on particular occasions. Exploring the Primal Health Research Database is a typical curiosity-enhancing situation. When updating our database, I constantly pause to consider plausible interpretations of correlations provided by epidemiologists. This enhanced curiosity is pushing us to be aware of data provided by a great variety of fast-developing scientific disciplines, so I want to offer a quick review of what everybody needs to know about emerging scientific disciplines such as epigenetics, modern bacteriology, the science of informational substances and their receptors and evolutionary biology.

Epigenetics

When the human genome project was declared complete in 2003, the DNA sequences were described as 'the recipe for making a person'.[1] Today the advent of epigenetics is at the root of a comprehensive rethinking of what makes two individuals different. This emerging discipline is based on the concept of gene expression. Some genes may be allocated a kind of label (an 'epigenetic marker') that makes them silent without altering the DNA sequences. This marker can be a DNA methylation, or a change in the nuclear protein content, such as modifications to histones. DNA methylation, which reduces gene expression, is the best-studied epigenetic modification, mainly because tools have been available to study it. The phenomenon of gene expression is influenced by environmental factors, particularly during the primal period. In relation to primal health research, epigenetics has a huge explanatory power; it gives a renewed importance to the concept of critical periods of development, and constitutes previously missing links between genetics, disease and the environment.

We already have enough data at our disposal to claim that, among humans, the period of birth is one of intense epigenetic activity, influenced by the mode of birth. A first Swedish study compared DNA methylation in white blood cells after pre-labour caesarean and after vaginal birth.[2] Another study provided similar results by looking at hematopoietic stem cell epigenetics.[3] Neither of these studies considered the case of birth by in-labour caesarean. A British study came to the conclusion that perinatal epigenetic analysis may be useful for identifying individual vulnerability to later obesity.[4]

Since it now appears that epigenetic markers (the 'epigenome') may be, to a certain extent, transmitted to subsequent generations, we should pay renewed attention to studies exploring the transgenerational effects of what happens during the primal period, and include them in our database.

We should not ignore the limits of what we can learn from the epigenetic perspective. We still have a lot to learn about long-lasting epigenetic differences related to mode of delivery.

Twenty-first-century bacteriology

When referring to modern bacteriology, the word 'revolution' is not too strong. The 'microbiome revolution' is a consequence of technological advances. As long as bacteriologists could only look at microscopes and cultivate microbes on Petri dishes, they could not see the 'unseen majority', since the growth conditions of many microbes are unknown. The turning point came when bacteriologists could dramatically expand their horizons thanks to the power of computer processing and new DNA sequencing technologies. For those who take the Primal Health Research Database as a point of departure, modern microbiology is apparently endowed with an enormous explanatory power.

Homo sapiens can now be viewed as an ecosystem, with a constant interaction between the trillions of cells that are the products of our genes (the 'host') and the hundreds of trillions of microorganisms that colonize the body (the 'microbiome'). The microbiome revolution is at the root of a real rethinking of the immune system, which needs specific stimulation to develop properly. We are in the process of learning that our health and our behaviour are highly influenced by our microbiome. The immune functions of the main components of the human microbiome (particularly the gut flora and the skin flora) have remained ignored until recently. Today it is acceptable to claim, for example, that our gut flora is responsible for about 80 per cent of our immune system.

Our current knowledge can be simplified by saying that to be born is to enter the world of microbes, and that the human microbiome is largely established in the perinatal period. This is confirmed by an accumulation of data regarding the gut flora[5,6], the skin microbiome,[7] the mouth microbiome,[8] and the milk microbiome.[9] The immunological perspective confirms that the perinatal period is a vital initial phase of interactions between host and microbiome. It can be seen as critical for immune programming.[10,11] For example, a comparative study of 15 infants born vaginally and nine infants born by caesarean section followed up from the age of one week until the age of two years confirmed the different immunological responses between the two groups. The differences were related to a lower total microbial diversity among those born by caesarean.[12]

In order to introduce these new issues to the framework of a modern lifestyle, we need to recall that until recently human babies were born in a bacteriologically familiar environment via the bacteriologically rich perineal zone. The term 'familiar' applies to the mother and to the child since, compared with the placenta of other mammals, the human placenta is highly effective at transferring the maternal antibodies called IgG to the fetus.[13,14,15]

Today, simplified safe techniques of caesarean section have brought radical changes in the way human microbiomes are established. A caesarean birth is associated with obligatory microbial deprivation in the neonatal period. At a time when the immune system is compared to a sensory organ that needs specific stimulations during critical periods of development, there are reasons to expect significant changes in the comparative prevalence of pathological conditions, even if, later on in life, dietary intake influences the structure and activity of the human microbiome.[16]

Further, there are reasons to anticipate an increased prevalence of dysregulations of the immune system. The concept of dysregulation of the immune system first suggests allergic diseases. It is significant that all studies of atopic diseases and asthma included in the Primal Health Research Database provide concordant results, with caesarean birth as a risk factor.[17, 18, 19, 20]

Interestingly, there are published studies confirming that the perinatal period should be looked at from immuno-bacteriological perspectives in order to interpret data about risk factors for asthma and allergic diseases. For example, in a double-blinded, placebo-controlled study, Finnish researchers randomised 1,223 mothers with infants at high risk for allergy to receive a probiotic mixture or a placebo during the last month of pregnancy, and their infants to receive it from birth until the age of six months. It appeared that perinatal supplementation of probiotic bacteria to high-risk mothers and babies conferred protection only to caesarean-born children.[21] In the light of such studies, not only can we claim that a caesarean birth is a risk factor for asthma and allergic diseases, but we also have at our disposal valuable clues for interpreting correlations established by retrospective epidemiological studies and for thinking in terms of cause and effect. Since caesarean-born babies are often exposed to antibiotics in the perinatal period, let us mention that pre- and perinatal exposure to antibiotics has been established as an independent risk factor for asthma and allergic diseases and also as a risk factor for serious infections in infancy caused by antibiotic resistant bacteria.[22, 23, 24]

Auto-immune diseases are other common dys-regulations of the immune system. The pathogenesis of some of them is probably related to the way the microbiome is established in the perinatal period. There are, in particular, concordant data suggesting that caesarean section is associated with an increased risk of childhood-onset type 1 diabetes. [25, 26]

Furthermore we are also now in a position to offer plausible interpretations of the studies that show that caesarean births are risk factors for obesity in childhood and in adulthood,[27, 28, 29, 30] since alterations in the gut microbiome of obese adults (and of adults with type 2 diabetes) have been clearly demonstrated.[31, 32] According to a Danish study, individuals with a low microbial diversity are characterised by more marked overall adiposity and insulin resistance than other individuals, and also by amplified inflammatory responses.[33] In general, a low microbial diversity is pathogenic. We can mention, as an example, the low microbial diversity of colicky babies.[34] Let us repeat that low gut microbial diversity is the main characteristic of caesarean-born children.[11]

This simplified – even simplistic – way to look at childbirth from a bacteriological perspective has obvious limits. It cannot explain, for instance, the radical differences between pre-labour and in-labour caesareans. We have already mentioned studies of the milk microbiome. The significant differences are between the milk of mothers who gave birth by pre-labour caesarean and those who gave birth by in-labour caesarean or vaginally.[35] These results suggest that it is not the operation per se that can alter the microbial transmission process to milk; other factors must be considered. A Canadian study of the gut flora of four-month-old babies found similar differences.[36]

Bacteriological studies demonstrating radical differences between pre-labour and in-labour caesareans underline the complexity of these issues. There are several ways to penetrate this complexity. One is to reconsider the premise of the 'sterile womb'. In fact, it is simplistic to claim that to be born is to enter the world of microbes. According to recent studies, it appears that most infants might incorporate an initial microbiome *before* birth.[37] This has already been confirmed by studies of the placental microbiome.[38, 39, 40] An important point is that the placental microbiome profile is akin to the maternal oral microbiome. Another perspective is to focus on the important new concepts of 'gut flora maturation' and the 'dynamics of infant gut microbiota'. These offer measures of postnatal development and are at the roots of the Singapore studies. They help us to understand the multiplicity of factors influencing gut flora maturation. Finally, we need to reassess the explanatory power of modern bacteriology.

Informational substances and their receptors

The concept of interaction between informational substances and their receptors is a basic tenet of the biology of multicellular organisms in general, and animals in particular. These organisms cannot exist without communication between neighbouring and distant cells. The science of informational substances is developing at such a high speed that previous classifications must be reconsidered. Even terms such as 'hormones' and 'endocrine glands' have become confusing. Until recently a hormone was an informational substance released in the bloodstream by a limited number of well-identified endocrine glands: pineal gland, pituitary gland, thyroid gland, parathyroid gland, pancreas, adrenal gland, ovaries and testes. This was the realm of endocrinology. Today we know that the adipose tissue, the digestive tract and even the heart release informational substances into the bloodstream that are often called hormones. Furthermore, a substance such as oxytocin, for example, which was originally presented as a hormone released by the posterior pituitary gland, can also carry information between nerve cells and therefore be classified as a neurotransmitter.

One of the easiest ways to classify the hundreds of

informational substances that have already been identified is to take into account their chemical structures. Some of them are steroids, which means that the parent molecule is cholesterol. This group includes the corticosteroids, such as cortisol, released by the adrenal gland, and also sex hormones such as testosterone, estrogens and progesterone. Steroids act on receptors located in the nucleus of cells. All of them are hormones according to the old classification. Others, especially neurotransmitters and substances involved in immunity, are peptides (made from the association of amino acids). They act on receptors located on the surface of the cells. Some of them are very small molecules, made of one amino acid or a few. This is the case of neurotransmitters such as acetylcholine. Adrenaline and other catecholamines are derived from the amino acids phenylalanine and tyrosine. Other peptides are more complex and are made up of a greater number of amino acids. For example, oxytocin and vasopressin contain nine amino acids. These more complex peptides are still usually classified as hormones. As for the prostaglandins, they derive from very long chain polyunsaturated fatty acids.

There have always been strong links between the science of informational receptors and primal health research. The discovery of brain receptors for opiates and oxytocin occurred in the same phase of the history of sciences as the advent of the branch of epidemiology we call primal health research. When Candace Pert revealed, in 1973, that there are opiate receptors in the human brain, she opened the way to a generation of research that led to the discovery of 'endorphins'.[41] We eventually learn that during the hour following birth, both mother and baby are supposed to be under the effect of opiates.[42, 43, 44] The history of oxytocin ran in the other direction. In 1979 a historic experiment showed that oxytocin has behavioural effects. Prange and

Pedersen found that if oxytocin is injected directly into the brain of mammals, maternal behaviour is induced.[45] This experiment triggered a real explosion of research, and the role of oxytocin brain receptors was confirmed. Thanks to that research it became possible to suggest a mechanism for the concept of a critical period for mother-baby attachment that had been introduced by ethologists.

After this reminder of the twentieth-century history of the relationship between primal health research and the science of informational substances, we can illustrate the fast development of this scientific discipline by mentioning some very recent twenty-first-century findings. It is now established that there are melatonin receptors in the human uterine muscle and that oxytocin and melatonin (the 'darkness hormone') work together. Furthermore, it has been scientifically demonstrated that acute inhibition of melatonin release with light suppresses uterine contractions... important facts in the age of electricity.[46,47] Interestingly, it is the blue wavelength that is most effective in melatonin suppression: during labour, it is more rational to rely on candlelight than on the scialytic lamp of a delivery room!

Stress-deprived babies born by pre-labour caesarean have low levels of adiponectin in their blood. Adiponectin is an informational substance involved in the metabolism of lipids and carbohydrates.[48] We have already mentioned that stress-deprived babies do not release noradrenaline, and for that reason their sense of smell is not functional at birth.[49] It is highly probable that the same lack of stress influences the development of the system of neurotransmitters: remember the effects of oxytocin-induced uterine contractions on the expression of the neurotransmitter GABA.[50] We still have a lot to learn about the role played in brain development in the perinatal period of informational substances such as uncoupled protein 2. The only limits to the science of informational substances are inherent to its youth as a discipline.

Evolutionary biology

There are also strong links between evolutionary biology and primal health research. Both perspectives suggest reasons and provide tools for looking towards the future. Scientific disciplines that suggest interpretations of the epidemiological studies included in our database are expanding the field of evolutionary biology. They put an end to what has been the dominant paradigm since the middle of the twentieth century. This dominant paradigm – often called Neo-Darwinism – was limited to the effect of the integration of genetics and Darwinian evolution, after what had been called 'the modern synthesis'.[51] Genetic mutation was considered the ultimate source of variation within populations. Evolution of the species was understood as a very slow process, measured by mutation rates. Within this restrictive theoretical framework, there was a lack of curiosity (or blindness?) about possible fast and spectacular transformations of species. During the twentieth century – apart from the Russian biologist and agronomist Lysenko – scientists did not dare to challenge orthodox ways of thinking and suggest that the heritability of acquired characteristics might be possible.

However, it now appears that the mother transmits more than her genes to her offspring. It is well accepted today that the mother, to a certain extent, transfers the way her genes have been educated through experience to be active or silent (i.e. she transfers her epigenome). In other words, she may transfer some acquired traits. She also transfers, to a certain extent, her microbiome. Furthermore, among humans, the genetic material of the mitochondria is inherited from the mother only.

Mitochondria are the so-called 'powerhouses' of cells. They convert energy into forms that are usable by the cell.

This new understanding of the comparative importance of maternal and paternal inheritance may have a great variety of implications. Throughout the history of mankind, there have always been correlations between the way human procreation was understood and the relationship between men and women.

It seems that before the time when our ancestors could learn from the breeding of domesticated mammals – i.e. before the Neolithic revolution and the introduction of animal husbandry – the procreative role of males was misunderstood, underestimated and even ignored. This was probably related to the long interval in our species between sexual intercourse and childbirth. At that time, motherhood was symbolic of the mystery of life. This is how one can explain the intriguing phenomenon of 'Venus figurines'. Over a hundred such figurines are known. They are pre-Neolithic, with the oldest ones being dated to more than 35,000 years before our own era. They have been found in places as diverse as Japan, China, Siberia and Western Europe. They are made of materials as diverse as ivory, serpentine rock, ceramic, limestone or hematite, but in spite of all their differences, their similarities are striking. In general, they exaggerate the parts of the body related to pregnancy, childbirth and breastfeeding, particularly the hips, abdomen, breasts, thighs and vulva. The mystery of life was obviously related to the female principle in general, and to motherhood in particular. The universality of the Venus figurine phenomenon is important for all students of human nature. It is the point of departure for any study of the history of the relationship between men and women.

With the advent of animal husbandry, the procreative power of males became obvious. The man appeared as the one bringing the 'seeds'. The role of the woman was to be an incubator and the provider of food. This new dominant cultural conditioning led to social systems in which males are the primary authority, occupying the roles of political leadership, and where fathers hold authority over women and children. Patriarchy means 'the rule of the father'. The 'rule of the father' being established, the universal transcendent emotional states were channelled towards the concept of one God-Creator as a paternal figure. In general patriarchal societies are also patrilineal, meaning that the child is given the name of the father and inherits from the father.

A significant reevaluation of the comparative roles of men and women in procreation was associated with the advent of modern genetics in the middle of the twentieth century. The biological maternal and paternal inheritances were considered equal, with the child receiving 50 per cent of its genes from the mother and 50 per cent from the father. Is it by chance that the 'women's liberation movement' began in the 1960s? Our societies are no longer typically patriarchal, since many children are born outside marriage, many married women keep their maiden names, and some children are given their mother's family name.

What will happen next? We are now starting to realise that mothers transmit more than their genes, and that maternal and paternal inheritances are not equal. What will happen if the issue of the comparative power of men and women in procreation is definitively overshadowed by the power of modern medicine?

Evolutionary biology, like primal health research, is curiosity enhancing. It also has its limits. These limits will recede when the experts in the discipline start looking towards the future.

CHAPTER 6

Bridges between scientific perspectives

As an interdisciplinary student of human nature, I am aware of the dangers of being the prisoner of one particular perspective. After reviewing the knowledge provided by a certain number of scientific disciplines and considering the limits of each of them, we must build as many bridges as possible between different ways of thinking. The point is to learn from a *combination* of data provided by a great variety of scientific teams. In an age of extreme scientific specialisation, it is more useful than ever to recall that 'The whole is greater than the sum of its parts'.

In order to illustrate the renewed importance of this old concept, we'll try to interpret some studies included in the Primal Health Research Database from as many perspectives as possible.

Autism

We have already looked at the example of the keyword 'autism' as a way to emphasise that concordant results may be obtained through a great variety of studies published over a period of several decades in a great variety of countries. We noticed that the results were concordant only in terms of timing: significant risk factors were identified in the perinatal period, and the nature of these risk factors was expressed differently according to the country, the research protocol, and the factors researchers had at their disposal. Among the most commonly mentioned risk factors were 'birth complications', 'birth asphyxia', 'caesarean', and 'low Apgar score'. It is notable that when researchers had at their disposal 'labour induction', it always appeared as a risk factor; as for 'labour augmentation', it appeared only once, but must be seriously considered because it was most relevant in the huge study from North Carolina.

Let us imagine that we are presenting these results to a great variety of authoritative scientific experts, and expecting them to suggest plausible interpretations.

First we meet a physiologist who has a special interest in the regulation of calcium metabolism. After looking attentively at the epidemiological studies, he explains that calcium homeostasis plays a fundamental role in the susceptibility to or development of some forms of autism. As a result he is not surprised that difficult births in general appear as risk factors, since abnormal calcium homeostasis in the uterus can result in poor quality uterine contractions.[1] According to him, the primary phenomenon is an alteration of uterine contractions in relation to disturbed calcium homeostasis.

Next we meet an expert in molecular psychiatry who has a special interest in epigenetics. He is knowledgeable about several studies which show that the period surrounding birth is a period of intense epigenetic activity;[2, 3] a short period during which a comparatively high number of genes are allocated a kind of label (a marker), making them silent. And he can report the results of his studies of 50 pairs of genetically identical twins, in which both twins were considered autistic or only one of them was. Among these 100 children, he could do an analysis of DNA methylation (as an epigenetic marker), an ideal method for identifying methylated regions in DNA among autistic subjects.[4] It is highly probable that the period surrounding birth has been critical in initiating this genetic activity in pairs of identical twins. The birth of one twin may be very different from the birth of the other one. From this perspective, we can conclude that activities in the period surrounding birth can explain the observed differences between genetically similar children.

In the age of the microbiome revolution, it is also essential to consider the point of view of an expert studying dysregulations of the immune system in relation to microbial deprivation in the neonatal period. There are reasons to look at autism from that standpoint. Our expert can explain in technical terms that autism is associated with an imbalance of Th1- and Th2-like cytokines.[5, 6, 7] The fact that mothers of autistic children are four times more likely to harbour anti-brain antibodies than unselected women of child-bearing age provides other reasons to look at autism in this framework.[8] It is also significant that maternal hypothyroidism – as an autoimmune disease – is a risk factor for the child becoming autistic.[9]

There are many kinds of experts in neurotransmitters, and all of them are highly specialised. However, let's present the epidemiological data to an expert in GABA. This expert is in a position to summarise the recently documented dysfunctions of the GABA system in autism – it has been demonstrated that oxytocin-induced uterine contractions are reversing the effects of GABA: this primary excitatory neurotransmitter becomes inhibitory. It is not surprising that some modern ways to interfere with the birth process can influence the prevalence of diseases characterised by a dysfunction of the GABA system.

Since the use of synthetic oxytocin is by far the most common medical intervention in childbirth, we must also give great importance to the point of view of an expert in oxytocin. This expert cannot help mentioning well-documented alterations in the oxytocin system of autistic children. Autistic children have comparatively low blood oxytocin levels.[10] While oxytocin levels increase with age in normal children, they remain low among those diagnosed autistic. Furthermore, autistic children have comparatively high blood concentrations of intermediates of oxytocin that probably accumulate due to incomplete processing.[11] According to this expert, at a time when most women receive synthetic oxytocin while giving birth, we can no longer deny the existence of problems arising from possible transfer via the placenta. Why does this remain an unexplored issue? Principally because oxytocin is not considered a 'real' medication, because chemically the synthetic form is no different from the natural hormone: it is a simple molecule. However, the equation is not simple, because the amount of oxytocin reaching the maternal blood stream via an intravenous drip is enormous compared to the amount of natural oxytocin the posterior pituitary gland can release. Furthermore, natural oxytocin is released through pulsations while synthetic oxytocin is delivered continuously. Another reason for ignoring the issue might be the discovery of enzymes that metabolise oxytocin in the placenta. This finding might have led to a hasty, tacit conclusion that synthetic oxytocin does not reach the baby.

Until now, there has been only one serious article published on this subject.[12] A team from Arkansas concluded that oxytocin crosses the placenta in both directions after measuring concentrations of oxytocin in maternal blood, in the blood of the umbilical vein and umbilical arteries, and also after perfusions of placental cotyledons. More precisely, the permeability is higher in the maternal-to-fetal direction than in the reverse. Eighty per cent of the blood reaching the fetus via the umbilical vein goes directly to the inferior vena cava via the ductus venosus, bypassing the liver, and therefore reaches the fetal brain immediately: it is all the more direct since the shunts (foramen ovale and ductus arteriosus) are not yet closed.

Since there is a high probability that a significant amount of synthetic oxytocin can reach the fetal brain, we must investigate the permeability of the so-called blood-brain barrier at this phase of human development. This 'barrier' restricts the diffusion of microscopic particles, including bacteria, and molecules such as oxytocin. However, Australian researchers presented evidence that the developing brain is more permeable to small lipid-insoluble molecules and that specific mechanisms, such as those involved in transfer of amino acids, develop gradually as the brain grows.[13] There is an accumulation of data suggesting that the blood brain barrier works in a specific way during fetal life.[14,15,16,17] Furthermore, it appears that its permeability can increase under the influence of oxidative stress,[18,19,20,21] which is common when a synthetic oxytocin drip is administered during labor. According to this expert, it is probable that, at a global level, we routinely interfere with the development of the oxytocin system of human beings at a critical phase for gene-environment interaction. It is therefore probable that there is a link between the increased incidence of disorders associated with documented alterations of the oxytocin system (such as autism[22, 23] and anorexia nervosa[24]) and the widespread use of oxytocin in obstetrics.

After listening to the scientists and accumulating theories, we now present the epidemiological data to practitioners who treat autistic children. A new, unexpected method recently became topical. The objective is to correct brain wave patterns utilizing Magnetic Resonance Therapy. We are told that this method is effective in treating autism and the kind of post-traumatic stress disorder that is common among war veterans. Is autism a variant of post-traumatic stress disorder? We have learnt from our multidisciplinary enquiry that it is easy to propose theories.

Obesity

Obesity is a condition that is now very prevalent in developed countries. Exploring the Primal Health Research database reveals many studies relating obesity in childhood and adulthood to the mode of birth. There is an obvious major difference between the studies about autism and those about obesity. We have already mentioned that, as early as 2006, after a series of important studies published at the beginning of the twenty-first century, it was already clear that mode of birth is an important factor in the genesis of autism, while it was difficult to detect risk factors during the period of conception, during fetal life, and after the birth. The case of obesity is radically different. Valuable studies exploring possible correlations between the mode of birth and the risks of obesity have only been published since 2010, although the keyword 'obesity' had been important since the origin of the database.

One of the oldest and most valuable studies in the database is about risk factors for obesity. It was published in 1976.[25] From October 1944 to May 1945, an acute famine affected the western Netherlands. The authors combined information about prenatal and early postnatal status at the time of the famine, with the weight and height of 300,000 men, 19 years later, being examined for military service. The main conclusion was that deprivation during the first half of pregnancy was related to significantly higher obesity rates at age 19, while deprivation during the last trimester of pregnancy and the first months after birth, was associated with lower obesity rates.

This historical study opened the way for further research about the long-term effects of intrauterine life during the Dutch famine. In one of these studies, the authors measured the body size of 741 people born at term between November 1943 and February 1947 in Amsterdam.[26] They compared people exposed to famine in late, mid, or early gestation with those born before, or conceived after, the famine period. It appeared that maternal malnutrition during early gestation was associated with higher Body Mass Index and waist circumference in 50-year-old women, but not in men. Another study looked at the glucose tolerance of adults who had been exposed either to the famine during fetal life, or who were born in the same area the year before the famine, or who had been conceived after the famine.[27] Glucose tolerance was significantly decreased among adults whose intrauterine life was during the period of starvation. The siege of Leningrad also exposed the entire population of a well-defined area to a severe famine. Among those exposed to malnutrition the prevalence of obesity and high blood pressure was stronger.[28]

Many other studies – particularly by David Barker's team in Southampton – researched the risks for obesity and related diseases in relation to birth weight and confirmed the results of the Dutch studies. Smoking in pregnancy[29, 30] was always found to be a risk factor, as were the long-term effects of certain drugs given to pregnant women. For example, it appears from one study that betamethasone given to a pregnant woman in order to prevent neonatal respiratory distress might result in insulin resistance in her child 30 years later.[31]

There have also been studies evaluating the prevalence of obesity in childhood, adolescence and adulthood in relation to infant feeding. In general, it seems that breastfeeding has a protective effect.[32, 33] However, the associations between breastfeeding, its duration, and the risks of being overweight in childhood, adolescence and adulthood have not been confirmed by large authoritative studies extending into adulthood such as the 1958 British birth cohort.[34] It appears from several of these studies that the weight of the mother is a stronger predictor of obesity than the mode of infant feeding. Such data suggest that the metabolic profile of a pregnant woman has more long-term influence than the kind of food consumed in infancy.

Interestingly, from studies that focus on the first week following birth, we could already conclude that the period between birth and age eight days is a critical window for nutritional programming. One study looked at the weight gain during this critical period of adults aged 20 and 32 who had been bottle-fed.[35] Another examined the first week of extra-uterine life of children of diabetic mothers.[36]

We emphasised the concordance between the results of all autism studies, whatever the research protocol, the country and the decade. It is rather different in the case of obesity (and overweight). For example, among studies about the risks in adulthood, one authoritative Brazilian study published in 2011 found that a caesarean birth is a highly significant risk factor for obesity at age 23–25, but a year later, the same authoritative medical journal published another Brazilian study according to which a caesarean birth is not a significant risk factor for obesity in childhood, adolescence and early childhood.[37, 38] There are also disagreements between studies of obesity in childhood. A German study revealed a greater likelihood of obesity at age two but not at age 6 or 10.[39] An American study examined the relationship between caesarean birth and the risk of overweight and obesity at age 12. They provided the results according to sex. It appeared that a caesarean birth is a risk factor for obesity and overweight among boys. Girls had an increased risk only for being overweight.[40] The Avon Longitudinal Study of Parents and Children followed up more than 10,000 subjects born in 1991–92 up to 15 years of age. When the sample was stratified by maternal weight, the risk of obesity was strong among children born of overweight or obese women. In contrast, the risk was not significant among children born of normal-weight mothers.[41] Only one study contrasted pre-labour and in-labour caesareans. Interestingly this issue did not appear in the abstract, so it has remained unnoticed.[42] It appeared that only 'unplanned' caesarean section (in practice that means 'in-labour') was clearly associated with a higher risk of obesity at age 3.

In spite of this lack of concordance between the results of studies found in our database, we have accumulated a sufficient amount of data to conclude that, in general, a caesarean birth is associated with a tendency to become overweight or obese later on in life. Our curiosity is thus stimulated. We have good reasons to present the data to scientific experts who might suggest interpretations.

Since adipose tissue is now considered an endocrine gland, releasing hormone-like substances, we'll first contact an expert in the informational substance called adiponectin. Adiponectin is involved in the metabolism of lipids and carbohydrates. The expert explains that the adiponectin found in high concentration in the cord blood does not have a maternal or placental origin: it is attributed to fetal fatty tissues.[43] He clarifies that babies need the stress of labour to release adiponectin: those born by pre-labour caesarean have comparatively low levels.[44] This difference from all other babies is confirmed by the fact that after in-labour caesarean the concentration of adiponectin in the milk is comparatively high.[45] Bringing these data together, the expert can conclude that, theoretically, babies born by pre-labour caesarean are more at risk than others of becoming overweight or obese.

The stomach is now also considered a kind of endocrine gland. This is a reason to interview an expert in the gastric hormone ghrelin, an appetite stimulant. We are told of a paradox. Obesity is commonly accompanied by increased hunger. However, the blood concentrations of ghrelin, the principal hunger hormone, are not elevated in obese people. The point is that their ghrelin system is dysregulated, so food intake is more efficiently stimulated in them than in lean humans. The mechanism is immunologic: in the case of obese people, ghrelin degradation is inhibited by specific antibodies.[46] These dysregulations probably start at birth, since the mode of birth is the only known factor influencing ghrelin levels in newborns.[47] Our expert in ghrelin, who is also an immunologist, emphasises that inflammatory phenomena are crucial in obesity and that, in the future, it is plausible that obesity will be treated with drugs targeting the immune system.[48]

In the age of the microbiome revolution, we cannot do without the point of view of an expert in the effects of microbial deprivation in the perinatal period. According to this expert, there is no need for long explanations. A breakthrough article in *Nature* identified the gut flora as a contributing factor to the genesis of obesity. Microbial populations in the gut are different between obese and lean people: obesity is associated with the same lack of microbial diversity as among babies born by caesarean; it is associated with the same changes in the relative abundance of the two dominant bacterial divisions (the Bacteroidetes and the Firmicutes).[49] We do not doubt that such changes have started at birth.

Interdisciplinarity

The exploration of the Primal Health Research Database via the keywords 'autism' and 'obesity' has offered an opportunity to show that by thinking across disciplines we prepare the ground for a new awareness. Curiosity and critical thinking strengthen interdisciplinary approaches. How can we develop these two human characteristics?

CHAPTER 7

Meanwhile

While waiting for a new awareness to take hold, we must be realistic and practical. The caesarean section is the most common operation worldwide. We must adapt to the current situation. How can we minimise the effects of stress deprivation and microbe deprivation that are experienced by many newborn babies?

Timing of the operation

The best way to minimise the effects of stress deprivation is undoubtedly to prefer, whenever possible, an in-labour caesarean. I know from personal experience that it is often possible to plan this kind of c-section. For example, breech presentation at term, is a situation in which, according to the usual guidelines, the vaginal route is considered too risky. How can we reconcile the point of view of babies, who adapt easily to 'in-labour non-emergency caesareans', and the point of view of adults who may prefer to control the timing of the operation?

We must also keep in mind that it is extremely difficult to predict what will happen after labour has begun. Even the most carefully planned scenarios can take surprising turns. I attended a woman who had had an operation to correct an abnormality of her cervix. It had been reconstructed and was made of hard, sclerous tissue. When she came to our hospital for her first delivery, her powerful contractions were ineffective: the only realistic solution was a caesarean section. Two years later she was pregnant again. Since we knew that in her particular case cervical dilation was impossible, we planned an in-labour caesarean section. She arrived at the hospital on the due date, in the middle of the night, after spending an hour in the car. All of us were ready for a second caesarean. However, the first child had been breastfeeding continuously on the journey to the hospital. When they arrived the mother's cervix was almost completely dilated, and the baby was born vaginally.

Immediate skin-to-skin contact

In the 1970s, scientists popularised 'bonding', and identified the first hour following birth as critical for mother and baby attachment. Since that time, it has gradually become more and more common to promote skin-to-skin contact between mother and baby immediately after a caesarean birth, on the operating table.[1, 2, 3] This has been the origin of such bizarre terms as 'natural caesarean' in the UK, or 'gentle caesarean' on the other side of the Atlantic. The rationale for these well-intentioned methods was originally based on a lack of understanding of the concept of the critical period for attachment. The short period following birth is deemed critical because it takes about an hour, for both the mother and the baby, to eliminate the hormones released during labour and delivery. Each of these hormones has a specific role to play in the bonding process. In the case of a caesarean birth, particularly a pre-labour caesarean, mother and baby are not in such a specific hormonal balance. The first hour following birth cannot, therefore, be considered really critical in terms of mother and baby attachment. In addition, during a caesarean, drugs classified as 'sympathomimetics' usually affect the mother's hormonal profile, in total contrast to when there is no interference with the physiological processes.

Although the original theoretical bases of immediate skin-to-skin contact after a caesarean are disputable, we must first emphasise that there are no known adverse outcomes. Of course, some difficulties must be overcome, in order to protect the baby on the operating table against noise, light and inappropriate temperature. Levels of melatonin (the darkness hormone) have been found to be low in the cord blood after a caesarean, and melatonin is protective because of its antioxidative properties.[4] There have been reassuring data about the stability of the infant's temperature, even in the thermal conditions of an operating room.[5] The satisfaction of the mothers must also be taken into consideration. This is why there is no reason to stop this practice. What is urgent is to introduce these issues in a renewed scientific context.

Entering the world of microbes

Today the main practical questions must be raised from immunological and bacteriological perspectives. How do we compensate for the effects of microbial deprivation after a birth by caesarean in a bacteriologically poor and unfamiliar environment, perhaps with occasional exposure to antibiotics? How can we facilitate the initial phases of an appropriate education of the immune system?

We must start with simple and realistic solutions, based on the fact that the human placenta is highly effective at transferring antibodies called IgG. The levels of IgG of a newborn baby born at term are at least 100 per cent of the maternal levels. This implies that, in general, microbes familiar to the mother are also familiar, and consequently friendly, to the baby. It is therefore easy to understand that, ideally, microbes familiar to the mother should be the first to colonise the baby's body. Since the extent of skin-to-skin contact between mother and baby after a caesarean is limited, which person is, bacteriologically speaking, the best substitute for the mother? If the parents live and sleep together then the baby's father is the ideal person. It is also easy, immediately after birth, to wrap the baby in clothes recently wore by the mother.

The gauze-in-the-vagina technique, evaluated by Maria Gloria Dominguez Bello in Puerto Rico, is based on a simple idea. A gauze pad is placed in the vagina in order to collect bacteria-laden secretions. Then, right after the caesarean birth, the baby's skin and mouth are gently swabbed. This technique is based on the assumption that human babies have been programmed for immediate colonisation by vaginal microbes. We have good reasons to challenge this assumption. I know from experience that when a birth is not socialised, when there have been no vaginal examinations during pregnancy and labour, and when the conditions for a real fetus ejection reflex are met, a 'birth in the caul' is comparatively common. When the bag of water remains intact, the first microbes met by the baby do not belong to the vaginal flora. We have good reasons to believe that birth in the caul was common before the socialisation of childbirth.

In all societies, there has always been a special interest in 'caulbearers'. They are supposed to be destined for lives of good health, fame and fortune. There are impressive lists of famous people born in the caul. Is it not paradoxical that until now being born protected against vaginal microbes was auspicious, and that now doctors have decided to treat this as a kind of microbial deprivation?

The Finnish study we have mentioned about the use of probiotics offers a rational and apparently effective way to overcome the plausible negative consequences of being born in a bacteriologically poor and unfamiliar environment. The researchers looked at mothers with infants at high risk for allergy. The mothers received a probiotic mixture or a placebo during the last month of pregnancy and their infants received it from birth until age six months. It appeared that supplementation with probiotic bacteria in high-risk mothers and children conferred protection only to caesarean-born children.[6] Why do we not try to replicate this unique study?

The main specific dysregulations of the immune system of caesarean-born children might inspire other strategies to minimise the effects of microbial deprivation. The main component of these dysregulations, as demonstrated at the age of two years, is easily expressed in technical terms: a reduced 'Th1 response'. These dysregulations are related to a lower microbial diversity and particularly to a lower abundance and diversity of certain groups of bacteria in the gut.[7] It appears that the mycobacteria of the 'Bacillus Calmette-Guérin' (commonly known as the tuberculosis vaccine) are potent inducers of the Th-1 response. It is, therefore, plausible that very early immunisation by BCG would be an easy, fast and cheap way to direct the early development of the immune system. Prospective studies evaluating the value of such a strategy should be ethically acceptable, since BCG has been routinely given in the neonatal period for many decades in a great variety of countries. What would be new would be to give it immediately after birth. The well-documented complications – particularly BCG osteitis – are exceptionally rare, and the outcomes are favourable with an adequate treatment.[8] One must keep in mind that all published non-specific effects on health of BCG are beneficial, particularly the effects on all-cause infant and childhood mortality in developing and developed countries.[9, 10] We must also save from oblivion an unexpected finding in a follow-up study covering the period between 1948 and 1998, which included American Indians and Alaska Natives who participated in a placebo-controlled BCG vaccine trial from 1935 to 1938. Data from 1,483 participants in the BCG vaccine group and 1,309 in the placebo group were analysed.[11] In the full text of the article a short sentence mentioned that the prevalence of type 2 diabetes was lower in the BCG group. This finding remained mysterious until recently. However, even before the 'microbiome revolution' I had suggested that BCG might have a future as a way to correct specific dysregulations of the immune system.[12]

Ironies

Our modern view of birth gives rise to many ironies. When all human beings, all over the world, were born vaginally in non-clinical environments, many mothers and newborn babies were ritually separated at birth, and the initiation of breastfeeding was delayed. Despite this history, today, as caesarean births are becoming more common, scientists have introduced the concepts of bonding, attachment and immediate skin-to-skin contact. Now that many babies are born in an operating room through an abdominal operation, medical staff are trying to establish sophisticated strategies to combine the rules of sterile surgical techniques and immediate skin-to-skin contact.

It is also ironic that birth with intact membranes has been considered the guarantee of good health and good

From Plato to Kerstin Uvnäs Moberg

For thousands of years students of human nature have been intrigued by the universal need to play. What is the meaning of the pleasure associated with an activity that has no immediate practical objective? As usual, when studying human nature, we must start with Plato, who claimed 'you can discover more about a person in an hour of play than in a year of conversation'. Then we can jump to the eighteenth century, when Jean-Jacques Rousseau wrote about the importance of observing play as a vehicle to learn about and understand children.

In the context of the 20th century, Friedrich Frobel, the author of *The Education of Man* (1903), was undoubtedly a pioneer: he came to the conclusion that 'play is the highest development in childhood, for it alone is the free expression of what is in the child's soul'. Maria Montessori, as an observer of childhood behaviour, had a strong interest in how children play; she noticed in particular that a child will repeat activities over and over until an 'inner need' is fulfilled. The concept of 'inner need' provides food for thought. Interestingly, Rudolf Steiner associated the concept of play and the concept of health: 'To a healthy child, playing is not only a pleasurable pastime, but also an absolutely serious activity. Play flows in real earnest out of the child's entire organism'. From the beginning of the history of psychoanalysis, a paramount importance was given to play. It started with Sigmund Freud, who wrote about the case of five-year-old 'Little Hans'. It culminated with Melanie Klein, who believed that child's play was essentially the same as free association used with adults. It became topical during the second half of the twentieth century after the publication of *Playing and Reality*, the well-known book by English paediatrician and psychoanalyst Donald Winnicott.

As an independent interdisciplinary researcher, and after spending a great part of his life focusing on the meaning of play and its formative function, Johan Huizinga, originally a Dutch historian, dared to introduce the term *Homo ludens* when referring to our species.

The 21st century will probably be characterised by a new generation of research. We are in an age of fast-developing scientific disciplines that will shed new light on the main characteristics of the genus *Homo*.

Since play is first and foremost a rewarding activity, modern physiologists are studying the major neurochemical pathway of the brain reward system. They are studying informational substances involved in pleasure. Today the focus is on oxytocin. The work of the Swedish physiologist Kerstin Uvnäs Moberg and colleagues is perhaps symbolic of the advent of a new generation of research. They found that dogs increase their levels of oxytocin and reduce their levels of cortisol when interacting with their owner. This is an opportunity to recall that play is not only a basic human behaviour. It is shared with high-achieving mammals, including all primates. Because play has often been associated with childhood, it is also worth recalling that human beings need to play at any age. Through ultrasound scans we know that in the middle of fetal life, a boy playing with his thumb and sucking it may have an erection and therefore be under the effect of oxytocin. We also know, on the other hand, that a centenarian may be happy to play cards with his pals.

The primal health research perspective

Will it be possible in the near future to study play from a primal health research perspective? It might be easier than to study the development of more vague – although essential – basic human characteristics. Curiosity, creativity and enthusiasm, in particular, which have strong links with play, cannot easily be studied from an objective scientific perspective.

Because drawing and painting are universal ways to play, they allow for objective scientific observation. Drawing as a way to play was well understood by Winnicott, whose name is associated with the famous 'squiggle game' with children.

I am convinced that the work of Arno Stern is currently the most promising way to study play from a primal health research perspective. Arno Stern's life and work are inextricably linked. He was born in Germany in 1924, where he spent the first part of his childhood, until the advent of Nazism in 1933. Then his family migrated to France. During World War Two, when Arno was a teenager, he left the occupied French zone to go to the south of France and ended up spending several years in a Swiss camp. This explains why, after the war, as a young adult, Arno had no professional qualifications. In 1946, at the age of 22, he found a job in a Parisian institution for war orphans. He got the children painting, and immediately understood how primordial the role of play is to this activity, for which he created an original environment. Thus, for nearly seven decades, Arno observed 'le jeu de peindre' ('the painting game').

The first lesson from the work of Arno Stern is that children's creations are as rich and personal as possible when the child feels completely free and is not given any suggested theme. The paintings made in the environments offered to children (and adults) are not destined to be seen, nor to be commented on by others. Arno kept all of them. The second lesson is that the capacity to draw appears very early among the gestures of a young child and is related to motor development. It follows a programmed process and does not develop through observation in particular environments.

In order to enlarge on what he was learning in Paris, Arno spent time with indigenous people in Mauritania, Peru, Niger, Mexico, Afghanistan, Ethiopia, Guatemala, and New Guinea. He had a special interest in paintings by children and adults who had never gone to school.

Stern is now summarising what he has learnt by using the concept of 'formulation' as a coherent and universal system. It accompanies the life of all human beings throughout all phases of their existence, regardless of their cultural conditioning. 'Formulation' is driven by inner necessity. It is a deep-rooted aspect of human nature.

The work of Arno Stern has been influential in some pedagogic circles. It has obviously inspired prominent French psychoanalysts such as Françoise Dolto.

In terms of the potential for research, the work of Arno Stern is unique. Tens of thousands of spontaneous paintings, from a great variety of cultural milieus, are now stored in electronic form. As birth is the formative phase of modern life that has been the most dramatically modified in recent decades, it should be possible to find out whether extreme deviations from its physiology might alter the need to draw and the way people draw spontaneously. For example, is the need and the capacity to draw altered, one way or another, among the generation born by pre-labour caesarean or after hours of synthetic oxytocin drips? Are there more children, nowadays, who don't feel the need to draw? Or, are there more children who have an insatiable need to play by drawing or painting, as if they would be more hard-wired for novelty seeking than children of previous generations? Is it possible that these personality traits are occasionally classified, today, as pathological and are associated with labels such as 'Attention Deficit Hyperactivity Disorder' (ADHD)?

Thanks to Stern's work, it is possible to study *Homo ludens* from a primal health research perspective. However it is significant that we cannot rely on his 'disciples'. When referring to the thousands of people he has inspired, Arno does not use this term, since disciples are supposed to follow the teaching of a master. Arno is humble enough to use the term '*Servant du jeu de peindre*'.

Considering play as a basic behavioural pattern gives us opportunities to notice that certain subjects can suddenly enter the field of 'serious' scientific research. We might offer other examples. Laura Uplinger, as an avant-garde interdisciplinary student of human nature and event organiser, surprised many 'serious' people by introducing in the programme of an international conference the unexpected theme of 'The Functions of Joy in Pregnancy'. Why should this be such a surprise?

CHAPTER 9

At the edge of the precipice

When considering the current phase of the history of childbirth and the history of mankind, some analogies may reflect a little dose of optimism while others are more pessimistic. I have occasionally used the phrase 'at the bottom of the abyss'. Such an easily understood analogy is appropriate to transmit pure pessimism and hopelessness. Since emerging scientific disciplines keep providing reasons for hope and therefore optimism, we'll adopt the phrase 'at the edge of the precipice'. When at the edge of a precipice it may still be possible to stop and take another direction. The best way to justify our choice is to refer once more to one of the most important scientific discoveries of the second half of the twentieth century, because this discovery has the power to challenge thousands of years of cultural conditioning and provides reasons for hope.

A twentieth-century scientific discovery

Those of my generation are in a position to recall how we suddenly learnt – from a scientific perspective – that a human newborn baby needs its mother. This had been understood for thousands of years – by farmers in particular – in the case of non-human mammals. However, even after the advent of the concept of evolution of the species, our deep-rooted cultural conditioning made it difficult for us to accept our mammalian roots. When I was an *externe* (medical student with minor clinical responsibilities) in a Paris hospital, in 1953, I had never heard of a mother who would have said, just after giving birth: 'Can I keep my baby with me?' The cultural conditioning was too strong. Everybody was convinced that the newborn baby urgently needed 'care' given by a person other than the mother. The midwife was quick to separate mother and baby by cutting the umbilical cord and putting the baby in the hands of a nurse. This is what she had learnt to do in midwifery school. At that time, the same would happen at home births. Then, while staying in the maternity unit, babies were in nurseries and mothers elsewhere. Mothers did not ask to stay in the same room as their baby.

The turning point had its roots in the middle of the twentieth century, when Konrad Lorenz, Niko Tinbergen and Karl Von Frisch collaborated in developing ethology as a separate sub-discipline of biology. Ethologists study a particular behavioural process rather than a particular animal group. For example, they may study aggression in a number of unrelated animals, including humans. It is easy to understand their capacity to inspire human studies. This is how, when they introduced the concept of critical periods for mother-newborn attachment, explaining that among mammals in general there is a crucial short period of time immediately after birth that will never happen again, they inspired studies of humans.

The time was ripe to evaluate the effects of immediate skin-to-skin contact between mother and newborn baby as an absolutely new intervention among humans. The names of Marshall Klaus and John Kennell in the United States are associated with such studies,[1] which were also conducted in Sweden.[2,3,4] In parallel, other researchers were interested in the behavioural effects of hormones that fluctuate in the perinatal period, particularly oestrogens.[5,6,7] This was also the decade when a sudden interest in the content of human colostrum developed. Until that time 'colostrum' was a fruitful keyword in veterinary medicine, but not in human medicine. In the 1970s, the focus was on local antibodies (IgA) and anti-infectious substances.[8,9,10] After thousands of years of negative connotations human colostrum was officially recognised as a precious substance.

In the 1970s we also learned that, when there is a free, undisturbed and unguided interaction between mother and newborn baby during the hour following birth, it is highly probable that the baby will find the breast during that hour: human babies usually express the 'rooting reflex' (searching for the nipple) during the hour following birth, at a time when the mother is still in a special hormonal balance and has the capacity to behave in an instinctive 'mammalian' way. The result of the complementary behaviour between mother and newborn baby is an early initiation of breastfeeding.[11, 12] For obvious reasons, nobody knew, before the 1970s, that the human baby was programmed to find the breast during the hour following birth.

The 1970s was also a period of rapid development in immunology and bacteriology. We mentioned the importance of studies about the easy and effective transfer of maternal antibodies (IgG) across the human placenta.[13, 14] This implies that the microbes familiar to the mother are also familiar, and therefore friendly, to the germ-free newborn baby. This was the beginning of a new vision of human birth from a bacteriological perspective. We were in a position to understand that the main questions are about the first microbes that occupy the territory and become the rulers of the territory. In other words, we were in a position to understand that from immunological and bacteriological perspectives a newborn baby ideally needs to be in urgent contact with the only person with whom he or she shares antibodies, and in a place that is bacteriologically familiar.

Immediate practical implications

This is how, in a nutshell, scientists revealed in the 1970s that a newborn baby needs its mother, challenging thousands of years of tradition. The discovery had immediate practical implications. It is not by chance that the concept of 'rooming-in' suddenly developed: it implies that mother and baby are in the same room while staying in the maternity unit. In fact, this concept originally appeared in some American hospitals during World War Two. But at that time the main reasons for this new organisational arrangement were practical; it was a way to adapt to the nursing shortage. In some hospitals, it has been a way to reduce the effects of epidemics of neonatal infections in nurseries. This concept spread out all over the world. In 1981, I was invited to speak at a conference about 'rooming-in' in Olomouc, Czechoslovakia. At that time, in the context of the communist regime, being critical of the concentrations of neonates in nurseries was considered revolutionary. Also in 1981 I went to Bogotà, Colombia, where they had been practising 'kangaroo care' for premature babies for two years (although the term 'kangaroo care' was not yet used). They had understood that when a premature baby does not need help to breathe the mother can be the best possible incubator.

Also at that time, there were unexpected effects from the sudden popularisation of the concept of 'bonding' immediately after birth. While scientific disciplines were studying the interaction between mother and newborn baby, some birth activists and theoreticians understood that the newborn baby needs its *parents*. It was exactly at that time that the idea of the baby's father participating in the birth began to spread all over the world. Today, immediate contact between mother and newborn baby, even on the operating table after a caesarean birth, is another practical consequence of the discoveries of the 1970s.

A transitory phase in our history?

In spite of these concrete practical changes, these scientific advances had limited effects. We can easily understand why: it is simplistic to claim that, according to modern scientific knowledge, a newborn baby needs its mother. In fact we should emphasise that the newborn baby ideally needs a mother who is in a specific physiological state just after giving birth, before hormones such as oxytocin, prolactin, endorphins, vasopressin and so on are eliminated. Ideally the birth would have occurred in such a context that the mother could start transmitting her microbiome to the baby. Finally, in our societies, these conditions are almost never met. Let us hope that we are in a transitory historical phase, between the time when the basic needs of newborn babies have been seriously identified and the time when we'll eventually dare to reconsider our understanding of the basic needs of labouring women.

CHAPTER 10

The gaps between science and tradition

Every day new scientific data challenge our cultural conditioning. While raising questions about midwifery in the past, present and future, we have had many opportunities to underline the contrast between science and cultural conditioning.

Countless examples

The way in which the basic needs of newborn babies are understood is a typical example. There are even cases of absolute contradiction, as in the case of the value of human colostrum. In many societies, colostrum was considered so harmful that it was expressed and discarded, yet in the current scientific context it is considered precious.

According to our dominant cultural conditioning, microbes are enemies we must kill. We associate microbes with diseases. When I was a medical student in the 1950s, one of the main preoccupations of the midwives was to protect the newborn baby against all microbes:

the labouring woman was given an enema, shaved and, after the birth, an antiseptic solution was spread on her nipples. Today the focus is on our symbiotic relationship with the world of microbes and the concept of microbial deprivation, particularly in the context of modern obstetrics. In the same vein, according to our cultural conditioning, stressful situations are considered negative and avoided. Today we are learning about the 'need' for stress. We need opportunities to release stress hormones, and to understand the possible negative effects of stress deprivation in situations such a birth by pre-labour caesarean.

At a time when, at a planetary level, the number of women who give birth without any pharmacological or operative assistance is becoming insignificant, it is urgent that we focus on the crucial divergence between two ways of understanding the birth process. Once more we must contrast the perspective of modern physiology on the one hand with, on the other hand, the deep-rooted bases of our cultural conditioning.

We already hinted at these two opposite ways of thinking when considering possible reactions to the question 'Do we need midwives?' While our question is deemed ludicrous in the framework of our cultural conditioning, it may be considered sensible in the twenty-first-century scientific context.

From a physiological perspective, the birth process is under the control of archaic brain structures shared – with some variants – by all mammals. From this starting point we can present the birth process as an involuntary process, and immediately refer to the main particularity of *Homo sapiens*, the enormous development of the part of the brain called the neocortex. To simplify, let us call this 'new brain' the thinking brain, or the brain of the intellect. Human instinctive behaviours that have mysteriously grown weaker or even disappeared are in fact repressed, or stifled, by neocortical activity. The concept of neocortical inhibition is key to understanding human nature.

The neo-cortex must stop working during the birth process. Giving birth is not the business of the brain of the intellect. When our neocortex is at rest we have more similarities to other mammals. During an easy, unmedicated birth, there is a time when the labouring woman seems cut off from our world, indifferent to what is happening around her. She tends to forget what she read, what she learnt and what her plans were. She may behave in a way that would usually be considered unacceptable in a civilised woman: for example, she might scream or swear (I have heard of labouring women who, during easy and fast births, have bitten the midwife: an excellent sign of reduced neocortical control!) She may be impolite. She may talk nonsense. She may find herself in the most bizarre, unexpected postures. These postures are often primitive, quadrupedal. When the labouring woman behaves as if 'on another planet', it means that there is reduced neocortical activity.

Starting from the concept of neocortical inhibition one can understand that the birth process needs to be protected against all stimulants of the neocortex – particularly light – and all attention-enhancing situations, such as being exposed to language, feeling observed, or perceiving a possible danger. The keyword is *protection*.

An unexpected way to learn about the crucial divergence

I find it useful to recall how, in 1964, I suddenly and unexpectedly understood the most important aspect of birth physiology. A friend of mine, a medical doctor working for a French pharmaceutical firm, gave me some samples of the recently synthesised Gamma-Hydroxybutyric acid (GHB).

It was the 1960s and he was already in a position to explain that it was an analogue of GABA (Gamma-Aminobutyric acid) and that it could not be dangerous since it was an integral part of the mammalian central nervous system.[1,2] This newly commercialised substance was presented as a sedative medication and as a promising agent in anaesthesiology. My friend added that, according to several preliminary studies, it shared the properties of oxytocin.

I assisted births with drips of Gamma-OH, as it was called in France. With these drips, labouring women would get completely crazy, shout in the corridors, pull out their intravenous needle, scare the midwife ... but the baby was born right away. Of course, such scenes were unacceptable in a hospital setting, and we had to be cautious about possible unreported negative side-effects. The main result of this audacious experiment – that we had to stop immediately – was a sort of revelation. I had understood that, when the activity of the neocortex is eliminated, human beings have more similarities with other mammals: this is what makes birth easy. I had understood the concept of neo-cortical inhibition and nature's solution for overcoming the human handicap. I had understood that GHB did not work like oxytocin, but facilitated its release. The important point was to realise that the neocortex of a labouring woman must not be stimulated.

Today GHB is not used as an anaesthetic agent. It has limited use in psychiatry. However, a widespread interest in this drug has developed recently, because it has become a major recreational drug and public health problem. Its ability to neutralise neocortical inhibitions explains the notoriety of this 'date-rape drug' – in other words, it is a compound used to facilitate sexual assault.

Since the 1960s we have known that GABA is the chief inhibitory neurotransmitter in the mature brain. Recently, considerable efforts have been devoted to examining the synergy between the 'darkness hormone' (melatonin) and the inhibitory agent GABA. Strong links between the two have already been documented. When science takes an an interest in the basic needs of labouring women, and when the concept of neocortical inhibition is applied to the birth process, studies of the triad melatonin-GABA-oxytocin will be considered essential.

Since the 1960s, my understanding of the effects of neo-cortical inhibition in childbirth has occasionally been reactivated. Once a young mother was celebrating the birth of her one-day-old baby in a double-occupancy hospital room. Her neighbour was in pre-labour. Glasses of champagne were exchanged. The effect of champagne was so spectacular that a baby was born through a 'fetus ejection reflex' on the way to the birthing room.[3] It is well known that champagne is a special wine. Thanks to the bubbles, alcohol is immediately brought to the brain. The ability of alcohol to change human consciousness has been known for ages ('*in vino veritas*'). Today we understand how alcohol works: one of its effects is to bind to the GABA receptors.[4] Furthermore, small amount of GHB have been found in wine.

I also learnt from the easy way schizophrenic women used to give birth before the widespread adoption of powerful antipsychotic treatments. It has been demonstrated that unmedicated schizophrenics have neocortical inhibition deficits. Interestingly, powerful antipsychotic drugs such as clozapine tend to interfere with the effects of GABA.[5]

It is also significant that psychedelic drugs used for their spiritual virtues have also been used to facilitate labour. This is the case of cannabis, which has been, and still is, a holy plant in many cultures all over the world. It is the most popular illegal drug worldwide. Although some European countries, Canada, Israel, and some US states have legalised medical cannabis, it is only through anecdotes and word of mouth that we have learnt about its actual effects in the particular case of the birth process. The biochemical effects of the cannabinoids – the most prevalent psychoactive substances in cannabis – have been widely studied. In 1990, the discovery of cannabinoid receptors located throughout the brain and body, along with endogenous cannabinoid neurotransmitters, suggested that cannabis affects the brain in the same manner as a naturally occurring brain chemical. Cannabinoids play an easily observed role in neocortical activity; they distort the perception of time and space. Furthermore, they affect pain transmission by interacting with the system of endorphins.[6] Their effects on the birth process can therefore be easily interpreted.

The daime, a drink known generically as ayahuasca, is another typical example of a drug used for both its spiritual virtues and its reputation for facilitating the birth process. It is the basis of a spiritual practice, the Santo Daime, founded in the Brazilian Amazonian state of Acre in the 1930s, which became a worldwide movement in the 1990s. As the daime is legal for religious purposes in Brazil, some midwives know the effects of this drug during labour and do not hesitate to report their observations. The decoction is made from two or more plants, such as the leaves of *Psychotria viridis*, which have high concentrations of the psychoactive compound dimethyltryptamine. Not only is this substance found in many plants, but it is also created in small amounts by the human body during normal metabolism. Its natural function remains undetermined. The stomach normally digests it, so it does not reach the brain if consumed orally, except when mixed with a 'monoamine oxidase inhibitor'

The still-dominant paradigm

A way of thinking in tune with physiological perspectives represents a typical paradigm shift. It is easy to contrast the two ways of understanding the birth process.

Our dominant cultural conditioning may be interpreted as the aftermath of thousands of years of beliefs and rituals. The basis of this deep-rooted conditioning is that a woman does not have the power to give birth by herself. It is the 'helping-supporting-coaching-managing paradigm'. One of the best ways to explore our cultural conditioning is to consider the way we speak and the roots of the words. If we ask a young mother who delivered her baby, we don't expect to hear her answer: 'I gave birth'. The Latin roots of the word 'obstetrics' imply that somebody must stand in front of the woman giving birth. We might offer countless other examples in many kinds of languages.

Many documented perinatal rituals, in a great variety of societies, imply that an agent of the cultural milieu must be present to 'do' something. Cord-cutting before the delivery of the placenta is a typical example. The main characteristics of this dominant cultural conditioning are shared by all literate post-agricultural societies on all continents. In Mexico, for example, I had the opportunity to visit a '*temazcal*', originally a sweat lodge in pre-Columbian societies. It was a small circular dome made of volcanic rock. Heated volcanic stones produced the heat. It was traditionally a place to give birth. My immediate comment was that such a small, round, dark and warm place was an ideal environment for making birth as easy as possible. However, in one of the written descriptions of birth I read, I found the phrase that in such a place 'the midwife can do certain things so that the birth is easier'. The focus was on what the midwife could do...

It is important to realise how our deep-rooted cultural

conditioning has been dramatically reinforced since the middle of the twentieth century. This has happened due to the effects of theories, such as the theories of Pavlov based on conditioned response, which are, directly or indirectly, at the root of all modern schools of 'natural childbirth'. It started with the term 'psychoprophylaxis'. It went on with the association of the term 'natural childbirth' with the names of doctors who described 'methods' or 'techniques', as if these two words were compatible with the word 'natural'. We might claim that the highest degree of cultural conditioning has been reached in the United States, with the term 'husband-coached childbirth'. It suggests that to give birth a woman needs to be guided by an expert and that the ideal expert is a man who, for obvious reasons, has no personal experience of childbirth: the coach just needs to be previously trained.

In the age of videos and photos the dominant cultural conditioning has also been reinforced by powerful visual messages. The usual cliché about 'natural childbirth' is a woman in a birthing pool surrounded by several people. This scenario, as a project, is a recipe for long and difficult labour. It is commonplace to associate 'water birth' and 'natural childbirth'. Let us recall that seals, which are much more aquatic than humans, go on to dry land to give birth! We originally introduced the concept of birthing pools in hospitals in order to avoid pharmacological assistance when the first stage of labour is very painful and difficult. It was useful, originally, to mention that 'a birth under water is bound to happen now and then'.[7] It was also essential to emphasise that, when in the birthing pool, the labouring woman should not feel observed.

The dominant conditioning related to childbirth really is cultural. Women, men, medical professionals and natural childbirth advocates are all subject to this conditioning.

This recent accumulation of contradictions between scientific knowledge and traditions is just one aspect of the crisis humanity finds itself in.

CHAPTER 11

In pain thou shalt bring forth children

The way we interpret labour pain is a basic aspect of our civilisation. In a famous ancient text summarising the main characteristics of human societies after the Neolithic crisis, the issue of labour pain is introduced in the framework of the domination of nature. We first read: 'Be fruitful, and multiply, and replenish the earth, and subdue it; and have dominion over the fish of the sea, and over the fowl of the air, and over every living thing that moveth upon the earth'. This is immediately followed by: 'In pain thou shalt bring forth children'. It appears that the domination of nature includes the socialisation of childbirth and its control by the cultural milieu. We have emphasised some of today's contradictions between scientific knowledge and cultural conditioning. However, we'll notice that, according to written tradition, there is an association between difficult births and 'the consumption of the fruit of the tree of knowledge'. This does not contradict the modern concept of neocortical inhibition.

Such considerations about labour pain offer the

opportunity to present an overview of current physiological perspectives. We learnt, in the late 1970s, that birthing mammals in general activate a physiological system of decreased perception of pain. The main component of this system is the release of opiate substances commonly called endorphins.[1, 2, 3, 4] The important point is that beta-endorphins (the main endorphins) play multiple roles, besides protection against pain. They induce, in particular, the release of prolactin.[5] We can therefore understand a chain of events: pain – endorphins – prolactin (involved in maternal behaviour and milk secretion). The lesson is that pain is a part of the physiological process: one cannot electively extract the pain and keep the other links of the chain of physiological events.

This is how one can interpret animal experiments suggesting a connection between the pain of birth and maternal behaviour. Nearly a century ago, Eugene Marais, in South Africa, observed that when birthing Kaffir buck mothers were given a few puffs of ether and chloroform, they refused to accept their newborn lambs.[6] More recently, Krehbiel and Poindron confirmed that after giving birth with epidural anaesthesia ewes do not take care of their lambs.[7] Among humans, of course, it is much more difficult to demonstrate the immediate effects, at an individual level, of removing the pain, for example with epidural anaesthesia. Among humans, who use sophisticated ways to communicate and create cultures, the collective dimension must be introduced: the questions must be raised in terms of civilisation.

When considering the particularities of childbirth in our species, we must once more focus on the huge development of the brain. We understand today that one of the roles of the brain, particularly the upper brain, is to 'interpret' signals coming from the body. The term 'upper brain', which is vague, might be more appropriate than the term 'neocortex' that we use in order to simplify. In the age of 'the neuromatrix theory of pain', it is acceptable to claim that 'pain is an output of the brain, not an input from the body'.[8] In this context it is easily understood that a reduction of neocortical activity is one of the main components of the physiological system of protection against labour pain. When the neo-cortex is at rest, peripheral messages are differently processed and don't have the same lasting effects. A reduced neo-cortical activity also implies the depression of memory. The depression of memory has obvious protective effects. It is also an effect of the well-known properties of opiates in general, and therefore of endorphins. Furthermore, the amnesic effects of oxytocin among humans are highly probable: memory tests have been used after a single dose of intranasal oxytocin.[9]

According to the social and cultural context, the brain interprets the potentially painful messages sent by the peripheral parts of the body differently. The signals associated with a broken nose during an official public boxing match are not processed in the same way if the same injury takes place at home during a common household activity. We assume that the signals sent by the labouring uterus to the brain were not originally interpreted in the same way as today. From what we know about childbirth before the Neolithic revolution, it appears that women knew to protect themselves against neocortical stimulations: they had a tendency to isolate themselves to give birth. After the advent of the domination of nature, and the control of childbirth by the cultural milieu, it became common for women to think 'labour is starting: time to call for help'. The socialisation of an event is a neocortical stimulant.

After the recent paradigm shift in brain sciences, questions must be phrased differently. It should not be:

'how to control labour pain?' But rather: 'how to make birth as easy as possible so that the physiological system of protection against pain is as effective as possible?' Finally, once more, we come to the issue of the basic needs of labouring women.

It is simplistic to separate advances in brain sciences and spectacular advances in our understanding of the transmission to the brain of signals from viscera such as the uterus. These advances are symbolised by the concept of the 'gate control theory of pain' that appeared in the 1960s. It is as if potentially painful stimuli from the peripheral parts of the body reach the spinal cord through a series of gates before passing towards the brain.[10] An important point is that these gates can be more or less open according to many factors. It appears, in particular, that skin stimulation at selective points can block the transmission of certain visceral painful stimuli. Since most women, in our societies, do not give birth in an appropriate environment, labour pain frequently reaches a pathological degree, with such a high level of stress hormones that the pain itself is an obstacle to the progress of labour. In such situations it makes sense, in order to break this vicious circle, to try controlling at the gate the transmission of stimuli sent by the uterus.

This leads me to recall the history of '*la réflexothérapie lombaire*'. In 1960 I learnt from an old surgeon a spectacular trick to treat acute kidney pains. It was the unilateral intracutaneous injections (one or two papules like nettle stings) of sterile water in the costomuscular angle, which is the depression just below the last rib. The injection is painful but it is a local pain that does not last more than a few seconds. The lumbar pain is eliminated immediately. Only some moderate anterior irradiations tend to persist. Some hours later it might be necessary to repeat the injections. This is how I thought of using such injections on both sides in obstetrics when cervical dilation does not progress and when the contractions are felt as lumbar pain. I found that when the lumbar pains are gone, only a discomfort above the pubic bone remains while cervical dilation is progressing. Originally I did not dare to write about this empirical method, avoiding the risk of being classified as a magician. Its mechanism was too mysterious. I became more audacious after hearing of the 'gate control theory of pain'. It suddenly became possible to explain how a superficial cutaneous pain could block, at the level of the spine, a painful message originating in a kidney or in the uterus. It is noticeable that the skin of the costomuscular angle is innervated by the posterior branch of the twelfth dorsal nerve, an appropriate zone, theoretically, to compete with such visceral pains. I was therefore in a position to publish a short paper in a French medical journal in 1975, about what I decided to call '*la réflexothérapie lombaire*' ('lumbar reflexotherapy').[11]

The technique spread to non-French-speaking countries through word of mouth. The term was lost in other languages and the site of injections in the lumbar region became vague. However, there were countless randomised controlled trials that confirmed the efficacy of this cheap technique. The *American Journal of Obstetrics and Gynecology* published a systematic review of 18 trials of any type of complementary and alternative therapies for labour pain.[12] All of these prospective, randomised controlled trials involved healthy pregnant women at term, and contained outcome measures of labour pain. They compared the effects of acupuncture, biofeedback, hypnosis, massage and respiratory autogenic training. Intracutaneous water injection was the only method that constantly appeared effective.

The immediate pain relief that is experienced following immersion in water at body temperature (37° Celsius) also needs to be discussed.[13] The effect is so fast when water is above the neutral temperature (i.e. above 33°C), that an action on the 'gates' is plausible through skin stimulation. The gate control theory of pain was refined in the 1980s, and we understand better how skin nerves transmitting heat can override the transmission of aching pains from visceral origin.

In spite of spectacular scientific advances the issue of labour pain remains eminently complex. It is a necessary perspective for all students of human nature.

CHAPTER 12

Will the symbiotic revolution take place?

This should become a central question encompassing all the others. We phrase it with an analogical hint to keep in mind that the extent of the responsibilities of our generation has suddenly reached a new order of magnitude. Some millennia ago, in the cradle of our western civilisations, at a time when Gaia was the personification of the Earth, serious people could not imagine more crucial questions than 'will the Trojan War take place?' Today, we are starting to raise questions about the relationship between humanity, Gaia and the survival of our species.

'Symbiosis' as the antithesis of 'Domination'

In order to explain the reason for the term 'symbiotic', we must refer once more to the spectacular crisis humanity went through with the advent of agriculture and animal husbandry. Few people had ever imagined, until the twentieth century, that the domination of nature that started thousands of years ago might suddenly reach its limits. However, we are now in a phase of our history when over-exploitation of the planet must urgently stop, since human actions already have an uncontrolled influence on many of the Earth's physical and biological processes, including human physiological processes. Humanity must invent new strategies for survival. We must prepare for a crisis, a revolution in the history of humanity and even in the history of 'Gaia'. And in order to qualify this revolution, a keyword is needed, one perceived as the antithesis of 'domination'. A possible relevant keyword is 'symbiosis'. Until now the prevalent way of thinking has led us to focus on competition, but we must now realise the importance of mutual assistance, not only among individuals within a species, but also *between* different species. The concept of symbiosis is key to realise how vital biodiversity is.

The new paradigm is already well advanced in biological circles. Variants of the term 'symbiosis' are more and more commonly used: 'endosymbiosis', 'ectosymbiosis', 'mutualism', 'commensalism' and so on. 'Symbiogenesis' appeared suddenly as an important mechanism of evolution. This new dominant way of thinking has been eloquently summarised by Dorion Sagan and Lynn Margulis: 'Life did not take over the globe by combat, but by networking'.[1] In order to realise the speed of this paradigm shift, we just need to recall that in 1976 the concept of 'selfish gene' was acceptable, although it conveys the idea that one gene can work in isolation... We must wonder how long it will take for the new way of thinking to go beyond specialised scientific circles.

Such a crisis implies a more symbiotic relationship between human groups. A unified humanity would make possible a 'dialogue' with Gaia. Such a crisis implies the need for new perspectives to explore the laws of nature in order to work with them, instead of controlling them. In practical words, it implies a more symbiotic relationship between our species and the ecosystems. Microorganisms are the foundations of all ecosystems. This is why the term symbiosis must also be understood at an individual level when considering plants and animals in general. While, until recently, our cultural conditioning presented microbes as enemies, *Homo sapiens* is now considered an ecosystem with a symbiotic relationship between our microbiome (the hundreds of trillions of microbes that live in our body) and our own cells. A lack of diversity within the human microbiome is pathogenic. If the revolution takes place, a new kind of relationship will be established between *Homo sapiens* and the world of microorganisms.

Reinventing Fire… and Birth

Theoretically, necessary solutions are already available in the current scientific context. 'Sustainability' is an emerging science born at the dawn of the twenty-first century. Hundreds of thousands of valuable articles about 'sustainable development' have been published by thousands of authors. The main obstacle is our cultural conditioning, which until now has a tendency to consider utopian all necessary solutions.

A typical example is offered by the use of fossil energy. In his book *Rediscovering Fire*, Amory Lovins, from the Rocky Mountain Institute (RMI), has designed an authoritative road map for a future without fossil fuels.[2] Several groups of scientists have devised budgets that would give the world a reasonable chance of limiting the rise in average global temperature to 2°C or less. However the International Energy Agency (IEA) predicts that, in spite of agreements between governments, the world will add one trillion tonnes of carbon dioxide to the atmosphere between now and 2040, using up the budgets established by the scientists.[3] Not only will the emissions of carbon dioxide inevitably increase, but also the emissions of methane, the second most prevalent greenhouse gas. Let us keep in mind that methane is the primary component of natural gas, and that it is emitted during the production, processing, storage, transmission and distribution of natural gas.

This discrepancy between knowledge and action should inspire urgent questions about the kind of *Homo* that has been dominating planet Earth for thousands of years.

We must take into account that the current generation of experts in sustainability look to the future as if the main characteristics of *Homo sapiens* were immutable. They ignore the fact that the dominant characteristics and personality traits of human beings, including respect for Gaia, might dramatically evolve in the near future. For example, experts in demography make projections without wondering what the fertility rates will be when pre-labour caesarean is the most common way to be born. In general, experts in sustainability don't raise questions about the development of respect for Mother Earth. The question is considered no more serious today than in 1979 when I published the book *Genèse de l'Homme écologique*.

Since the primary questions we raise are about rapid evolution of *Homo sapiens* in relation to the way babies are born, let's look at the difficulties experienced by those trying to 'reinvent birth'. We'll show the problematic links between the words 'necessary' and 'utopian' by analysing a birth scenario from the perspective of modern physiologists.

Let us imagine a labouring woman in a small, dark and warm room. There is nobody around, apart from one experienced and silent midwife sitting in a corner knitting. All the details of this apparently simple scenario can be interpreted from the perspective of modern physiology.

We mentioned that the midwife is knitting. It is more

than a detail.[4, 5] Studies have confirmed the physiological effects of repetitive tasks. At the April 2004 British Psychological Society conference, Emily Holmes, from Cambridge University, presented her studies of the effects of repetitive tasks, such as knitting, in stressful situations. She concluded that repetitive tasks are very effective in reducing tension. She also referred to the use of worry beads in many cultures, such as Greece, as a way to cope with stressful situations. The level of adrenaline released by the midwife in a birthing place is an important issue, since adrenaline is contagious and easily transmitted to the labouring woman. This can be demonstrated through studies of direct brain-to-brain communication, such as the exploration of the 'mirror neuron system'. It simply means that when we are close to a person who is in a specific emotional state, we activate the same part of our brain as this person. If the level of adrenaline of the midwife is high, it is probable that the labouring woman will increase her level of adrenaline, and therefore reduce her release of oxytocin: the birth will be difficult and longer. This is a good example of how sophisticated methods, such as methods of brain imaging, can help us to revive common sense.

We also mentioned that the labouring woman is in a small, dark room. Remember the importance of been protected against all stimulation of the neocortex, particularly attention-enhancing situations, such as feeling observed. The risk of feeling observed is reduced in a small, dark room. Melatonin – the 'darkness hormone' – and oxytocin work together. We must not ignore the word 'warm' either, since the level of adrenaline is higher when the ambient temperature is low. Even the word 'one' needs to be interpreted. The risks of feeling observed are minimised if there is nobody around, apart from 'one' midwife. Interestingly, this was understood in traditional societies. According to a Persian proverb, 'when there are two midwives, the baby's head is crooked',[6] while, according to a Hungarian saying, 'the baby is lost when there are two midwives'. Similar beliefs have been reported among South American indigenous ethnic groups. It is easy to understand that the labouring woman feels more secure (i.e. releasing a low level of adrenaline) with an experienced midwife and it is also easy to comment on the word 'silent', since language is the main stimulant of the neocortex. Finally, when describing this scenario as an example (rather than a model) of a situation usually compatible with a birth as easy as possible, we noticed that the midwife does not behave as an observer, staying in front of the labouring woman: she is sitting in a corner, keeping a low profile.

For many reasons, such a scenario appears as utopian, or culturally unacceptable, or simply unrealistic. It inspires reactions such as: 'To promote knitting is sexist' or 'What about the baby's father?' or 'What if the midwife is a man?'

Flirting with utopia

This illustrates the paradoxical situation in which we are today. It is culturally acceptable to claim that a control of the rates of caesareans is necessary. To claim that it is necessary to create conditions for as many women as possible around the world to give birth thanks to the release of a cocktail of love hormones is also culturally acceptable. Claiming that it is necessary to create conditions for the immune system of as many babies as possible to be educated, immediately after birth, by millions of microbes familiar to the mother is also culturally acceptable. In a word, the promotion of 'natural childbirth' is widespread and culturally acceptable. However, at the same time, describing the conditions needed to make a birth as easy as possible is culturally

CHAPTER 13

What is the sex of angels?

Considering that the symbiotic revolution might take place, thanks to the conceivable emergence of a spectacular new awareness, we must keep in mind our point of departure and reconsider our question about the need for midwives. Why this haunting feeling that perhaps we are wasting time in pointless arguments?

In order to go one step further, I suggest the analogy of the interminable high-level theological debates, some millennia ago, about the sex of angels. These enigmatic beings at the service of the Divine are mentioned in the Torah, the Bible ('angelos' are messengers), the intertestamental books (such as the Book of Enoch) and the Qur'an. It was, therefore, a serious and crucial issue in a particular cultural context.

The turning point in the debate occurred when two kinds of angels mentioned in the Bible – the rebels and the authentic ones – were discussed. The rebel angels became demons and tempted humans: while Incubi licentiously enter women's bodies, Succubi seduce men. The authentic angels practise the abstinence prescribed to humans who consecrate their lives to God: authentic angels embody total, asexual unity. From the time when such a dualism was revealed, the original question had to be rephrased.

Lesson from an analogy

This analogy is pertinent and helpful since we are reaching a point in the debate when it is inevitable to contrast two kinds of midwives. We have previously underlined that there are now two ways to understand the birth process. From the perspective of modern physiologists, birth appears as an involuntary process under the control of archaic brain structures. In general one does not try to help an involuntary process. However, it may be useful to be aware of possible inhibitory situations. In the case of the birth process, inhibitory factors are easily identified through such concepts as adrenaline-oxytocin antagonism and neocortical inhibition. This implies that the birth process must ideally be protected against any situation that might increase the level of adrenaline or stimulate the neocortex. In the paradigm introduced by physiologists, let us repeat that the keyword is 'protection'.

The other way to understand the birth process is the effect of thousands of years of cultural conditioning. It is the still-dominant 'helping-guiding-controlling-coaching-managing paradigm'. This paradigm, characterised by its disempowering vocabulary, is reaching limits beyond which midwives belong to an endangered species, even though the term can persist during a certain phase of history. If we keep going in the same direction, we'll reach the ultimate phase of socialisation of childbirth. There will be no more need for midwives. The need will be for medical teams managing perinatal situations. Within such teams, there will be experts designing protocols, updating these protocols according to authoritative published studies, and controlling the way they are followed. There will be

'laborists', in charge of labour management. There will be obstetrical anaesthesiologists. There will be technicians in charge of intravenous catheters, fetal monitoring devices and so on. There will be neonatologists. There will be, of course, laparotomists (surgeons opening the abdominal wall). In practice, laparotomists will be experts in caesarean section, since most abdominal operations will be performed via 'keyholes' or natural orifices.

But it is another matter altogether if we enter a new paradigm inspired by the perspective of modern physiology. Such a paradigm shift might be facilitated by events playing the roles of catalysts. For example, it is plausible that in the near future authoritative studies may provide concordant results regarding the long-term side effects of pharmacological assistance during labour, particularly the use of synthetic oxytocin. It is also plausible that specialised journals may offer overviews of the multiple negative side-effects of stress deprivation associated with pre-labour caesareans. New reasons to avoid both pre-labour births and caesarean births at the stage of real emergency are already appearing.[1,2] These are the conditions needed for the advent of simplified strategies in obstetrics, based on an improved understanding of the basic needs of labouring women in the age of simplified techniques of caesarean.

After the paradigm shift

If such a paradigm shift is possible, the midwife as a birth protector will play a central role. The basis of the strategy will be to favour two kinds of births: straightforward vaginal birth and in-labour non-emergency caesarean section. The prototypal scenario will be simple: its first objective is to create conditions for a birth as easy as possible. As long as the experienced midwife is optimistic, there is no reason to consider another option than the vaginal route. As soon as the midwife becomes pessimistic, it is probably wise to perform a caesarean section without waiting for an emergency. Of course, this prototypal scenario will need to be adapted to every particular case.

Simplified obstetrical strategies will give new importance to scores and tests that help providers to decide to perform a caesarean without delay, before an emergency arises. The birthing pool test is a typical example of a tool adapted to simplified strategies. It is based on a simple fact. When a woman in hard labour enters the birthing pool and is immersed in water at the temperature of the body, a spectacular progress in the dilation is supposed to occur within an hour or two.[3] If the already well-advanced dilation remains stable in spite of water immersion, privacy (no camera!), silence, dim light and no man around, one can conclude that there is a major obstacle. It is wiser to perform an in-labour non-emergency caesarean. I adopted this test after analysing the outcomes of the rare cases in which dilation had not progressed in the water. I realised that a caesarean had always been necessary in the end, more often than not after long and difficult first and second stages.[4]

The paradigm shift also implies that we need new ways to select candidates for midwifery schools. Personal experience of life and personality traits will be the main criteria. It is worth recalling that in the randomised controlled trial about 'doulas' in Houston, Texas, the prerequisite to participate in the study was a personal experience of 'a normal labour and vaginal delivery with a good outcome'.[5] It should become obvious that, in general, labouring women feel more secure when protected by a midwife who has a positive experience of giving birth. It should also become obvious that personality traits and empiric knowledge may be more important than a

scientific background. If the paradigm shift is delayed, it will become more and more difficult to find midwifery candidates with the appropriate profile, since the number of women who rely on their natural hormones to give birth to babies and placentas is becoming insignificant.

Once again, we must emphasise that today, whatever the topic, we constantly need words like 'limit', 'edge', 'frontier' or 'border'. When considering the history of childbirth, we came to the conclusion that we are at the edge of the precipice. When considering the effects of the Neolithic revolution, we came to the conclusion that we have reached the limits of the domination of nature. When discussing the concept of interdisciplinary perspective, we looked at the limits of what each emerging scientific discipline can offer. When observing that there are two kinds of midwives, we mentioned the limits of the 'helping-guiding-controlling-coaching-managing paradigm'.

When analysing possible solutions to the threats humanity currently faces, we are at the limits between what is realistic and what is utopian. Throughout this book, we have remained at the border of the Realm of Utopia. In this context, our precise question about the survival of midwifery must be understood as an introduction to the broader question: is the survival of humanity utopian?

ADDENDUM

Can humanity
survive medicine?*

During the academic year 1948–1949, I followed a program commonly called PCB (Physics, Chemistry and Biology) in Paris. It was usual for French students of my generation to spend a year in 'La Faculté des Sciences' between secondary school and medical school.

A premature question

During that year, I was fascinated by lectures given by Louis Gallien, a professor of 'animal biology'. Gallien was a respected expert in 'rational selection' of domesticated animals.[1] Once, in the middle of a lecture, he digressed into talking about *Homo sapiens*. With diplomacy, he expressed his worries about the time when medicine would become powerful enough to completely neutralise the laws of natural selection. This implied that medicine would have a 'dysgenic effect' by contributing to a gradual deterioration of the human gene pool. For obvious

---

* To be read after July 2030

reasons, our professor did not want to say more about *Homo sapiens*, and he never published the idea. In that phase of our history, it was morally risky to mention the laws of selection in the particular case of our species. The related topic of voluntary selection – eugenics as a social philosophy – was irrelevant and even taboo. Nobody, at that time, was worried about the power of medicine. The media had just reported the case of the first patient with rheumatoid arthritis to be treated with cortisone. Antibiotics were not yet widely used. Most surgeons did not dare to venture beyond the abdomen and the limbs. A caesarean section was a last-resort risky operation. There were no medically assisted conceptions.

However, the professor's brief digression was to be the highlight of my academic year. It often came back to my mind. I had realised that some spectacular changes in our lifestyles might unintentionally modify the characteristics of our species. Later on, I happened to read, in a paper published decades before the age of intercontinental flight, that 'the motorbus, by breaking up inbred village communities, was a powerful eugenic agent'.[2] And, in the scientific context of the 1970s, possible transformations of *Homo sapiens* in relation to modes of birth slowly began to dawn on me.

Neutralised laws of natural selection

The time has come to revisit the worries of Louis Gallien. We have suddenly reached the phase of history when medicine has effectively neutralised the laws of natural selection, thus entailing an obligatory dysgenic effect. It is a major turning point after nearly four billion years of the history of life. This turning point is spectacular when considering reproductive medicine. Until recently, as a general rule, only women who had a tendency to give birth easily had many children. Today, in an age of fast and easy techniques of caesarean, the number of children per woman is influenced by other factors than her capacity to give birth. Until recently, as a general rule, very fertile people had more children. Today a high degree of fertility is moderated by medicalised contraception, while infertility can be treated by medically assisted conceptions.

There are countless examples of diseases that medical advances have made compatible with reproduction. Adolescents and adults with such diseases now approach developmental milestones, including sexual and reproductive ones, like their healthy peers. This is true of type 1 diabetes, a disease with a strong genetic component. Today, children and adolescents diagnosed as diabetic can reach adulthood and have children. Women who have had a kidney failure treated by kidney transplant or dialysis can have babies. Teenagers who had cancer treated by chemotherapy can store their eggs or sperm and reproduce in later life. We must also mention purely genetic diseases such as cystic fibrosis, which is related to mutations in the gene for a protein called CFTR. Until now most people have had no copy of the mutated gene. Some people have only one copy of the mutated gene: they are carriers. People with cystic fibrosis have two copies. One in 30 Caucasian Americans is a carrier of a cystic fibrosis mutation. What will it be after many more generations? Not only can people with cystic fibrosis reach adulthood thanks to modern medicine, but their low degree of fertility can be treated. For example, men who reach adulthood but are not fertile because they don't have 'vas deferens' (ducts between the testes and the penis) can have children in medically assisted conceptions. It is still impossible to foresee to what extent, in the future, the dysgenic effect of medicine will be balanced by the eugenic effect of appropriate information, as in the case of a well-informed

person rationally avoiding having children with a carrier.

It would take volumes to consider all the possible effects of neutralising laws of natural selection among humans. But we can provide food for thought through an example. About 15 per cent of the European population lacks the main Rh antigen. In other words, a certain percentage of the population is Rh negative. Until the second half of the twentieth century, Rh negative women had a smaller than average number of surviving babies. Now fatal diseases related to Rh incompatibility are easily prevented. Rh negative women have the same average number of children as those who are Rh positive. It is easy to see that after a number of generations there will be a significantly increased proportion of the human population lacking the main Rh antigen, and therefore an increased proportion of Rh negative mothers carrying Rh positive babies.

A vicious circle

The neutralisation of the laws of natural selection is an undisputable fact. We are not in the realm of opinions. Basic mathematical analyses can easily reach the conclusion that, with the advent of increasingly effective medicine, a gradually increasing proportion of the human population will become more and more dependent on medicine. It should even be feasible, with the data we already have at our disposal, to design sophisticated computer programmes to foresee the time when such a high proportion of the human population is dependent on medicine that health budgets exceed all others. The dysgenic effect of medicine might become ruinously expensive.

When thinking long-term, we must keep in mind that the neutralisation of this basic law of life is a side-effect of medicine (including veterinary medicine). It does not directly involve the world of microorganisms. This is why *Homo* is put at a disadvantage in the war against microbes initiated by the advent of antibiotics. Today resistance of microbes to antibiotics – an effect of natural selection – is considered a major public health problem that very few people foresaw of in the middle of the twentieth century. Although the laws of natural selection have been understood for a long time, it remains difficult for our contemporaries to consider their enormous practical implications when our own species is involved. It is easier to include energy, pollution, climate change, water shortages, food and financial security among the major problems of our time and ignore the transformation of *Homo sapiens* in relation to radically new aspects of our lifestyles.

Genetically modified human beings

These difficulties are one of the reasons why I have always cautiously tried to postpone questions about the effects of medicine on natural selection. I found it more urgent to focus, in particular, on probable transformations of *Homo sapiens* in relation to the way babies are born.[3] It is plausible that these transformations will be detectable after a comparatively small number of generations of highly medicalised births. My original plan was to postpone premature questions until my one hundredth birthday, in July 2030. However, because I have doubts about whether I will be able to write at that time, I decided to write this preliminary draft. Before phrasing the right questions in a more definitive way, we should wait for some clarifications about the real potential of recent spectacular scientific and technical advances.

Gene therapy is among the recent technical advances that might oblige us to rephrase our questions. The

principle is to insert a gene into a patient's cells to treat or prevent disease. As long as the therapeutic genes are transferred into non-sex cells ('somatic gene therapy') one cannot expect spectacular effects on the following generations, since the modifications will not be inherited. It will be another matter when 'germline gene therapy' becomes ethically, technically and financially acceptable. If sperm or eggs are genetically modified, all the cells in the organism will contain the modified gene and the effect of the therapy will be heritable. We are reaching a phase in the history of sciences when it is theoretically possible to transform species by inheritable modifications of genomes! Shall we open the door to genetically modified humans?

Different orders of magnitude

In general, many of our contemporaries are ready to share preoccupations related to developments in genetics and reproductive technologies. It is plausible that, in spite of ethical and financial obstacles, more and more parents will be aware of the genetic material of their children long before birth. Let us just mention the techniques of pre-implantation genetic diagnosis and detection of free fetal DNA in maternal blood. It is easy to foresee the advent of a new era of eugenics. In this new framework, the decisions will not be made by governmental institutions, but by the family. What kinds of traits, apart from gender and the absence of genetic diseases, will be considered desirable by the parents? Intelligence? Artistic talents? Life expectancy? Body shape? Beauty? Can we imagine, for example, a time when it will be fashionable to select children whose genomes are compatible with a strong capacity for empathy, or a strong capacity to give birth (if the choice is to have a daughter)? Once more, we are flirting with Utopia!

As long as we focus on such spectacular aspects of reproductive medicine, we are not ready to balance the amplitude of its dysgenic and eugenic effects. The dysgenic effects are related to the mass practice of established modern medicine at a planetary level: the future of mankind is at stake. For many reasons, the eugenic effects remain comparatively occasional and limited. We are in a paradoxical situation. Because of the moral connotation associated with words, it is commonplace to be scared by the eugenic effects of medicine and, at the same time, to ignore its obligatory and vast dysgenic effects. These dysgenic effects will be amplified in the near future, when enormous financial resources are diverted towards the development of 'precision medicine', so that prevention and treatment strategies may be based on individual variability.

When comparing the eugenic and the dysgenic effects of modern medicine, the orders of magnitude are different. I assume this will be understandable in July 2030: 'Nothing is more powerful than an idea whose time has come'.

REFERENCES

CHAPTER 1

1. Schultz, A. *The Life of Primates*, Weidenfeld and Nicolson, London, 1969.
2. Trevatan, W.R. 'Fetal emergence patterns in evolutionary perspective', *American Anthropologist*, 1988; 90:674-681.
3. Rosenberg, K.R., Trevatan, W.R. 'Birth, obstetrics and human evolution', *BJOG* 2002; 109:1199-1206.
4. Engelmann, George J. *Labor Among Primitive People*. J.H. Chambers &Co., St Louis, 1884.

CHAPTER 2

1. Daniel Everett, *Don't sleep, there are snakes*, Profile Books, 2008.
2. Eaton, S.B., Shostak, M., Konner, M. *The paleolithic prescription*, Harper and Row, New York, 1988.
3. Marjorie Shostak, *Nisa*, Earthscan, London, 1990.
4. Wulf Schiefenhovel, *Childbirth among the Eipos*, New Guinea. Film presented at the Congress of Ethnomedicine, 1978, Gottingen, Germany.
5. Skoglund P, Mallick S, et al. Genetic evidence for two founding populations of the Americas. *Nature* 2015 July 21 .doi: 10.1038/nature14895
6. Monahan M, Boelaert K, Jolly K, et al. Costs and benefits of iodine supplementation for pregnant women in a mildly to moderately iodine-deficient population: a modelling analysis. *Lancet Diabetes Endocrinol.* 2015 Aug 7. pii: S2213-8587(15)00212-0. doi: 10.1016/S2213-8587(15)00212-0. [Epub ahead of print]
7. Carlson, S.E., Rhodes, P.G., et al. 'Effect of fish oil supplementation on the n-3 fatty acid content of red blood cell membranes in preterm infants', *Pediatric Research* 1987; 21(5):507-510.
8. Huerta-Sánchez, E., Jin, X., Asan, Bianba, Z., et al. 'Altitude adaptation in Tibetans caused by introgression of Denisovan-like DNA'. *Nature*. 2014 Jul 2. doi: 10.1038/nature13408. [Epub ahead of print]
9. *Science* http://dx.doi.org/10.1126/science.1250368, 2014
10. Newton, N., Foshee, D., Newton, M. 'Experimental inhibition

of labor through environmental disturbance', *Obstetrics and Gynecology* 1967; 371-377

11. Newton, N. 'The fetus ejection reflex revisited', *Birth* 1987; 14: 106-108.

12. Odent, M. 'The fetus ejection reflex', *Birth* 1987; 14: 104-105.

13. McGraw, M.B. 'Swimming Behavior of the Human Infant', *Journal of Pediatrics* 1939;15:485-90.

14. Endevelt-Shapira Y1, Shushan S2, Roth Y3, Sobel N4. 'Disinhibition of olfaction: Human olfactory performance improves following low levels of alcohol', *Behav Brain Res.* 2014 Jun 25;272C:66-74. doi: 10.1016/j.bbr.2014.06.024. [Epub ahead of print]

CHAPTER 3

1. Liu, S., Liston, R.M., Joseph, K.S., et al. 'Maternal mortality and severe morbidity associated with low-risk planned cesarean delivery versus planned vaginal delivery at term.' *CMAJ* 2007; 176(4):455-60.

2. Hankins, G.D., Clark, S.M., Munn, M.B. 'Cesarean section on request at 39 weeks: impact on shoulder dystocia, fetal trauma, neonatal encephalopathy, and intrauterine fetal demise.' *Semin Perinatol.* 2006 Oct; 30(5):276-87.

3. Liu, X., Landon, M.B., Cheng, W., Chen, Y. 'Cesarean delivery on maternal request in China: what are the risks and benefits?' *Am J Obstet Gynecol.* 2015 Jan 29. pii: S0002-9378(15)00099-X. doi: 10.1016/j.ajog.2015.01.043. [Epub ahead of print]

4. Li, H-T., Ye, R., Achenbach, T., Ren, A., Pei, L., Zheng, X., Liu, J-M. 'Caesarean delivery on maternal request and childhood psychopathology: a retrospective cohort study in China.' *BJOG* 2010; DOI: 10.1111/j.1471-0528.2010.02762.x.

5. Gonzales-Valenzuela, M.J., Garcia-Fortea, P., et al. 'Effects of oxytocin used during delivery on development: A retrospective cohort study.' *Journal of Clinical and Experimental Neuropsychology* 2014;11:1-11 DOI: 10.1080/13803395.2014.926864.

6. García-Fortea, P., González-Mesa, E., Blasco, M., Cazorla, O., Delgado-Ríos, M., González-Valenzuela, M.J. 'Oxytocin administered during labor and breast-feeding: a retrospective cohort study.' *J Matern Fetal Neonatal Med.* 2014 Jan 13. [Epub ahead of print].

7. Olza Fernández, I., Marín Gabriel, M., Malalana Martínez, A., et al. 'Newborn feeding behaviour depressed by intrapartum

oxytocin: a pilot study.' *Acta Paediatr.* 2012 Jul;101(7):749-54. doi: 10.1111/j.1651-2227.2012.02668.x. Epub 2012 Apr 4.

8. Bell, A.F., White-Traut, R., Rankin, K. 'Fetal exposure to synthetic oxytocin and the relationship with prefeeding cues within one hour post-birth.' *Early Hum Dev.* 2012 Oct 16. pii: S0378-3782(12)00239-3. doi: 10.1016/j.earlhumdev.2012.09.017. [Epub ahead of print]

9. Odent, M. 'Synthetic oxytocin and breastfeeding: reasons for testing an hypothesis.' *Medical Hypotheses* 2013;81 (5):889-891.

10. Odent, M. *Childbirth and the Future of Homo sapiens,* Pinter & Martin, London 2013.

11. Laughon, S.K., Branch, D.W., Beaver, J., Zhang, J. 'Changes in labor patterns over 50 years.' *Am J Obstet Gynecol.* 2012 May; 206(5):419.e1-9. Epub 2012 Mar 10.

12. Al-Mufti, R., McCarthy, A., Fisk, N.M. 'Survey of obstetricians' personal preference and discretionary practice.' *Eur J Obstet Gynecol Reprod Biol* 1997; 73:1-4.

13. Gabbe, S.G., Holzman, G.B. 'Obstetricians' choice of delivery.' *Lancet* 2001; 357: 722.

14. Rob Stein, 'Elective caesareans judged ethical', *The Washington Post,* October 31 2003. Page A32

15. Condon, J.C., Jeyasuria, P., Faust, J.M., Mendelson, C.R. 'Surfactant protein secreted by the maturing mouse fetal lung acts as a hormone that signals the initiation of parturition.' *Proc Natl Acad Sci USA.* 2004 Apr 6;101(14):4978-83. Epub 2004 Mar 25.

16. Hauth, J.C., Parker, C.R.Jr, MacDonald, P.C., Porter, J.C., Johnston, J.M. 'A role of fetal prolactin in lung maturation.' *Obstet Gynecol.* 1978 Jan;51(1):81-8.

17. Varendi, H., Porter, R.H., Winberg, J. 'The effect of labor on olfactory exposure learning within the first postnatal hour.' *Behav Neurosci.* 2002 Apr;116(2):206-11

18. Odent, M. 'The early expression of the rooting reflex.' Proceedings of the 5th International Congress of Psychosomatic Obstetrics and Gynaecology, Rome, 1977. London: Academic Press, 1977: 1117-19.

19. Odent, M. 'L'expression précoce du réflexe de fouissement.' In: *Les cahiers du nouveau-né* 1978; 1-2: 169-185

20. Hermansson, H. Hoppu, U., Isolauri, E. 'Elective Caesarean Section Is Associated with Low Adiponectin Levels in Cord Blood.' *Neonatology* 2014;105:172-174 (DOI:10.1159/000357178).

21. Cabrera-Rubio, R., Collado, M.C., Laitinen, K., et al. 'The human milk microbiome changes over lactation and is shaped by maternal weight and mode of delivery.' *Am J Clin Nutr.* 2012 Sep;96(3):544-51. doi: 10.3945/ajcn.112.037382. Epub 2012 Jul 25.

22. Azad, M.B., Konya, T., Maugham, H., et al. 'Gut microbiota of healthy Canadian infants: profiles by mode of delivery and infant diet at 4 months.' *CMAJ* February 11, 2013 cmaj.

23. Dogra, S., Sakwinska, O., Soh, S., Ngom-Bru, C., Brück, W.M., Berger, B., Brüssow, H., Lee, Y.S., Yap, F., Chong, Y., Godfrey, K.M., Holbrook, J.D. 2015. 'Dynamics of infant gut microbiota are influenced by delivery mode and gestational duration and are associated with subsequent adiposity.' *mBio* 6(1):e02419-14. doi:10.1128/mBio.02419-14.

24. Bagci, S., Berner, A.L., et al. 'Melatonin concentration in umbilical cord blood depends on mode of delivery.' *Early Human Development* 2012; 88(6):369-373.

25. Christensson, K., Siles, C., et al. 'Lower body temperature in infants delivered by caesarean section than in vaginally delivered infants.' *Acta Paediatr* 1993;82(2):128-31.

26. Simon-Areces, J., Dietrich, M.O., Hermes, G., et al. 'Ucp2 Induced by Natural Birth Regulates Neuronal Differentiation of the Hippocampus and Related Adult Behavior.' *PLoS ONE,* 2012; 7 (8): e42911 DOI: 10.1371/journal.pone.0042911

27. Tyzio, R., Cossart, R., Khalilov, I., Minlebaev, M., Hubner, C.A., Represa, A., Ben-Ari, Y, Khazipov R. 'Maternal oxytocin triggers a transient inhibitory switch in GABA signaling in the fetal brain during delivery.' *Science* 2006; 314: 1788-1792.

28. Downes, K.L., Hinkle, S.N., Sjaarda, L.A., et al. 'Prior Prelabor or Intrapartum Cesarean Delivery and Risk of Placenta Previa.'*American Journal of Obstetrics and Gynecology* 2015, http://www.ajog.org/article/S0002-9378(15)00005-8

29. Prior, E., Santhakumaran, S., Gale, S., et al. 'Breastfeeding after cesarean delivery: a systematic review and meta-analysis of world literature.' *Am J Clin Nutr.* 2012 May; 95(5):1113-35. doi: 10.3945/ajcn.111.030254. Epub 2012 Mar 28.

30. Zanardo, V., Savona, V., Cavallin, F., et al. 'Impaired lactation performance following elective delivery at term: role of maternal levels of cortisol and prolactin.' *J Matern Fetal Neonatal Med.* 2012 Sep;25(9):1595-8. doi: 10.3109/14767058.2011.648238. Epub 2012 Feb 6.

31. Levine, L.D., Sammel, M.D., Hirshberg, A., et al. 'Does stage of labor at time of cesarean affect risk of subsequent preterm birth?' *Am J Obstet Gynecol.* 2014 Sep 30. pii: S0002-9378(14)01020-5. doi: 10.1016/j.ajog.2014.09.035. [Epub ahead of print]

CHAPTER 4

1. Odent, M. 'Childbirth and Prevention of the Diseases of Civilisation.' *Nutrition and Health* (Berkhampsted, Hertfordshire).01/1983;1(3-4):161-164.
2. Hattori, R., Desimaru, M., Nagayama, I., Inoue, K. 'Autistic and developmental disorders after general anaesthetic delivery.' *Lancet* 1991; 337:1357-58.
3. Odent, M. 'Between circular and cul-de-sac epidemiology.' *Lancet* 2000; 355 (April 15): 1371.
4. Hultman, C., Sparen, P., Cnattingius, S. 'Perinatal risk factors for infantile autism.' *Epidemiology* 2002; 13: 417-23.
5. Zwaigenbaum, L., Szatmari, P., Jones, M.B., Bryson, S.E., MacLean, J.E., Mahoney, W.J., Bartolucci, G., Tuff, L. 'Pregnancy and birth complications in autism and liability to the broader autism phenotype.' *J Am Acad Child Adolesc Psychiatry.* 2002 May;41(5):572-9.
6. Glasson, E.J., Bower, C., Petterson, B., et al. 'Perinatal factors and the development of autism.' *Arch Gen Psychiatry* 2004; 61: 618-27.
7. Larsson, H.J., Eaton, W.W., Madsen, K.M., Vestergaard, M., Olesen, A.V., Agerbo, E., Schendel, D., Thorsen, P., Mortensen, P.B. 'Risk factors for autism: perinatal factors, parental psychiatric history, and socioeconomic status.' *Am J Epidemiol.* 2005 May 15;161(10):916-25; discussion 926-8
8. Stein, D., Weizman, A., Ring, A., Barak, Y. 'Obstetric complications in individuals diagnosed with autism and in healthy controls.' *Compr Psychiatry* 2006 Jan-Feb;47(1):69-75.
9. Duan, G., Chen, J., Yao, M., Ma, Y., Zhang, W. 'Perinatal and background risk factors for childhood autism in central China.' *Psychiatry Res.* 2014 Jun 24. pii: S0165-1781(14)00478-8. doi: 10.1016/j.psychres.2014.05.057. [Epub ahead of print]
10. Froehlich-Santino, W., Londono Tobon, A., Cleveland, S.,et al. 'Prenatal and perinatal risk factors in a twin study of autism spectrum disorders.' *J Psychiatr Res.* 2014 Jul;54:100-8. doi: 10.1016/j.jpsychires.2014.03.019. Epub 2014 Mar 29.
11. Gregory, S.G., Anthropolos, R., Osgood, C.E., et al. 'Association

of autism with induced or augmented childbirth in North Carolina Birth Record (1990–1998) and Education Research (1997-2007) databases.' *JAMA Pediatr.* 2013 Oct;167(10):959-66. doi: 10.1001/jamapediatrics.2013.2904.

12. Weisman O, Agerbo E, Carter CS, et al. Oxytocin-augmented labor and risk for autism in males. *Behav Brain Res.* 2015 May 1;284:207-12. doi: 10.1016/j.bbr.2015.02.028. Epub 2015 Feb 20.

13. Xiang AH, Wang X, Martinez MP, et al. Association of maternal diabetes with autism in offspring.*JAMA.* 2015;313(14):1425–1434pmid:25871668.

14. Mengying Li, M. Daniele Fallin, et al. The Association of Maternal Obesity and Diabetes With Autism and Other Developmental Disabilities. Pediatrics February 2016

15. Walker, C.K., Krakowiak, P., Baker, A., et al. 'Preeclampsia, Placental Insufficiency, and Autism Spectrum Disorder or Developmental Delay.' *JAMA Pediatr.* 2014 Dec 8. doi: 10.1001/jamapediatrics.2014.2645. [Epub ahead of print]

16. Lehti, V., Brown, A.S., Gissler, M. 'Autism spectrum disorders in IVF children: a national case-control study in Finland.' *Hum Reprod.* 2013 Mar; 28(3):812-8. doi: 10.1093/humrep/des430. Epub 2013 Jan 4.

17. Sandin, S., Nygren, K.G., Iliadou, A., et al. 'Autism and mental retardation among offspring born after in vitro fertilization.' *JAMA* 2013 Jul 3;310(1):75-84. doi: 10.1001/jama.2013.7222.

18. Padhye, U. 'Excess dietary iron is the root cause for increase in childhood autism and allergies.' *Med Hypotheses.* 2003 Aug;61(2):220-2.

19. Schmidt, R.J., Tancredi, D.J., Krakowiak, P., et al. 'Maternal Intake of Supplemental Iron and Risk of Autism Spectrum Disorder.' *Am J Epidemiol.* 2014 Sep 22. pii: kwu208. [Epub ahead of print]

20. Taylor, B., Miller, E., et al. 'Autism and measles, mumps, and rubella vaccine: no epidemiological evidence for a causal association.' *Lancet* 1999; 353: 2026-9.

21. Kaye, J.A., Melero-Montes, M., Jick, H. 'Mumps, measles, and rubella vaccine and the incidence of autism recorded by general practitioners: a time trend analysis.' *BMJ* 2001; 322: 460-3.

22. Dales, L., Hammer, S.J., Smith, N.J. 'Time trends in autism and in MMR immunization coverage in California.' *JAMA* 2001; 285 (9): 1183-5.

23. Madsen, K.M., Hviid, A., et al. 'A population-based study of

measles, mumps, and rubella vaccination and autism.' *N Engl J Med* 2002; 347(19): 1474-5.

24. Hviid, A., Stellfeld, M., Wohlfahrt, J., Melbye, M. 'Association between thimerosal-containing vaccine and autism.' *JAMA.* 2003 Oct 1;290(13):1763-6

25. Krehbiel, D., Poindron, P., et al. 'Peridural anaesthesia disturbs maternal behaviour in primiporous and multiporous parturient ewes.' *Physiology and behavior.* 1987; 40: 463-72.

26. Lundbland, E.G., Hodgen, G.D. 'Induction of maternal-infant bonding in rhesus and cynomolgus monkeys after caesarian delivery.' *Lab. Anim. Sci* 1980; 30: 913.

27. Boksa, P., Wilson, D., Rochford, J. 'Responses to stress and novelty in adult rats born vaginally, by cesarean section or by cesarean section with acute anoxia.' *Biol Neonate.* 1998;74(1):48-59.

28. El-Khodor, B.F., Boksa, P. 'Birth insult increases amphetamine-induced behavioral responses in the adult rat.' *Neuroscience.* 1998 Dec;87(4):893-904.

29. Black M, Battacharya S, et al. Planned cesarean delivery at term and adverse outcomes in childhood health. *JAMA* 2015;314(21):2271-2270.

30. Simon-Areces, J., Dietrich, M.O., Hermes, G., et al. 'Ucp2 Induced by Natural Birth Regulates Neuronal Differentiation of the Hippocampus and Related Adult Behavior.' *PLoS ONE,* 2012; 7 (8): e42911 DOI: 10.1371/journal.pone.0042911

CHAPTER 5

1. Alla Katsnelson, 'Epigenome effort makes its mark.' Published online | *Nature* 6 October 2010; 467, 646 | doi:10.1038/467646a

2. Schlinzig, T., Johansson, S., Gunnar, A., et al. 'Epigenetic modulation at birth – altered DNA-methylation in white blood cells after Caesarean section. *Acta Paediatr.* 2009;98:1096–9.

3. Almgren, M., Schlinzig, T., Gomez-Cabrero, D., et al. 'Cesarean delivery and hematopoietic stem cell epigenetics in the newborn infant: implications for future health?' *Am J Obstet Gynecol.* 2014 Jun 18. pii: S0002-9378(14)00465-7. doi: 10.1016/j.ajog.2014.05.014.

4. Godfrey, K.M., Sheppard, A., Gluckman, P.D., Lillycrop, K.A., Burdge, G.C., McLean, C., Rodford, J., Slater-Jefferies, J., Garratt, E., Crozier, S.R., Emerald, B.S., Gale, C.R., Inskip, H.M., Cooper, C., and Hanson, M.A. 'Epigenetic gene promoter methylation at

birth is associated with child's later adiposity.' *Diabetes* 60: doi: 10.2337/db10-0979 (2011)

5. Gronlund, M.M., Lehtonen, O.P., Eerola, E., Kero, P. 'Fecal microflora in healthy infants delivery.' *J Pediatr Gastroenterol Nutr* 1999; 28(1): 19-25.

6. Palmer, C., Bik, E.M., DiGiulio, D.B., Relman, D.A., Brown, P.O. 'Development of the human infant intestinal microbiota.' *PLoS Biol.* 2007 Jul;5(7):e177. Epub 2007 Jun 26.

7. Capone, K.A., Dowd, S.E., Stamatas, G.N., Nikolovski, J. 'Diversity of the human skin microbiome early in life.' *J Invest Dermatol* 2011 Oct;131(10):2026-32. doi: 10.1038/jid.2011.168. Epub 2011 Jun 23

8. Lif Holgerson, P., Harnevik, L., Hernell, O., Tanner, A.C., Johansson, I. 'Mode of birth delivery affects oral microbiota in infants.' *J Dent Res.* 2011 Oct;90(10):1183-8. Epub 2011 Aug 9.

9. Cabrera-Rubio, R., Collado, M.C., Laitinen, K., Salminen, S., Isolauri, E., Mira, A. 'The human milk microbiome changes over lactation and is shaped by maternal weight and mode of delivery.' *Am J Clin Nutr.* 2012 Sep;96(3):544-51. doi: 10.3945/ ajcn.112.037382. Epub 2012 Jul 25.

10. Molloy, E.J., O'Neill, A.J., Grantham, J.J., et al. 'Labor Promotes Neonatal Neutrophil Survival and Lipopolysaccharide Responsiveness.' *Pediatr Res* 2004 May 5.

11. Gronlund, M.M., Nuutila, J., Pelto, L., et al. 'Mode of delivery directs the phagocyte functions of infants for the first 6 months of life.' *Clin Exp Immunol* 1999; 116(3): 521-6.

12. Jakobsson, H.E., Abrahamsson, T.R., Jenmalm, M.C., et al. 'Decreased gut microbiota diversity, delayed Bacteroidetes colonisation and reduced Th1 responses in infants delivered by Caesarean section.' *Gut.* 2013 Aug 7. doi: 10.1136/ gutjnl-2012-303249. [Epub ahead of print]

13. Cederqvist, L.L., Ewool, L.C., Litwin, S.D. 'The effect of fetal age, birth weight, and sex on cord blood immunoglobulin values.' *Am J Obstet Gynecol.* 1978 Jul 1;131(5):520-5

14. Garty, B.Z., Ludomirsky, A., Danon, Y., Peter, J.B., Douglas, S.D. 'Placental transfer of immunoglobulin G subclasses.' *Clin Diagn Lab Immunol.* 1994 Nov;1(6):667-9.

15. Hashira, S., Okitsu-Negishi, S., Yoshino, K. 'Placental transfer of IgG subclasses in a Japanese population.' *Pediatr Int.* 2000 Aug;42(4):337-42.

16. David, L.A., Maurice, C.F., Carmody, R.N., et al. 'Diet rapidly

and reproducibly alters the human gut microbiome.' *Nature.* 2013 Dec 11. doi: 10.1038/nature12820. [Epub ahead of print]

17. Pistiner, M., Gold, D.R., Abdulkerim, H., Hoffman, E., Celedon, J.C. 'Birth by cesarean section, allergic rhinitis, and allergic sensitization among children with a parental history of atopy.' *J Allergy Clin Immunol.* 2008 Aug;122(2):274-9. Epub 2008 Jun 20

18. Bager, P., Wohlfahrt, J., Westergaard, T. 'Caesarean delivery and risk of atopy and allergic disease: meta-analyses.' *Clin Exp Allergy.* 2008. Apr;38(4):634-42.

19. Xu, B., Pekkanen, J., Hartikainen, A.L., Järvelin, M.R. 'Caesarean section and risk of asthma and allergy in adulthood.' *J Allergy Clin Immunol.* 2001 Apr;107(4):732-3.

20. Roduit, C., Scholtens, S., de Jongste, J.C., et al. 'Asthma at 8 years of age in children born by caesarean section.' *Thorax.* 2009 Feb;64(2):107-13. doi: 10.1136/thx.2008.100875. Epub 2008 Dec 3.

21. Kuitunen, M., Kukkonen, K., Juntunen-Backman, K., et al. 'Probiotics prevent IgE-associated allergy until age 5 years in cesarean-delivered children but not in the total cohort.' *J Allergy Clin Immunol.* 2009 Feb;123(2):335-41. doi: 10.1016/j.jaci.2008.11.019. Epub 2009 Jan 8.

22. Stensballe, L.G., Simonsen, J., Jensen, S.M., Bønnelykke, K., Bisgaard, H. 'Use of antibiotics during pregnancy increases the risk of asthma in early childhood.' *J Pediatr.* 2013 Apr;162(4):832-838.e3. doi: 10.1016/j.jpeds.2012.09.049. Epub 2012 Nov 6.

23. Jedrychowski, W., Gałaś, A., Whyatt, R., Perera, F. 'The prenatal use of antibiotics and the development of allergic disease in one year old infants. A preliminary study.' *Int J Occup Med Environ Health.* 2006;19(1):70-6.

24. Glasgow, T.S., Young, P.C., Wallin, J., et al. 'Association of intrapartum antibiotic exposure and late-onset serious bacterial infections in infants.' *Pediatrics.* 2005 Sep;116(3):696-702.

25. McKinney, P.A., Parslow, R., Gurney, K.A., Law, G.R., Bodansky, H.J., Williams, R. 'Perinatal and neonatal determinants of childhood type 1 diabetes. A case-control study in Yorkshire, U.K.' *Diabetes Care.* 1999 Jun;22(6):928-32.

26. Cardwell, C.R., Stene, L.C., Joner, G., et al. 'Caesarean section is associated with an increased risk of childhood-onset type 1 diabetes mellitus: a meta-analysis of observational studies.' *Diabetologia.* 2008 May;51(5):726-35. Epub 2008 Feb 22.

27. Huh, S.Y., Rifas-Shiman, S.L., Zera, C.A., et al. 'Delivery by

caesarean section and risk of obesity in preschool age children: a prospective cohort study.' *Arch Dis Child.* 2012 Jul;97(7):610-6. Epub 2012 May 23.

28. Blustein, J., Attina, T., Liu, M., et al. 'Association of caesarean delivery with child adiposity from age 6 weeks to 15 years.' *Int J Obes* (Lond). 2013 Jul;37(7):900-6. doi: 10.1038/ijo.2013.49. Epub 2013 Apr 8.

29. Goldani, H.A., Bettiol, H., Barbieri, M.A., et al. 'Cesarean delivery is associated with an increased risk of obesity in adulthood in a Brazilian birth cohort study.' *Am J Clin Nutr.* 2011 Jun;93(6):1344-7. doi: 10.3945/ajcn.110.010033. Epub 2011 Apr 20.

30. Darmasseelane, K., Hyde, M.J., Santhakumaran, S., Gale, C., Modi, N. 'Mode of delivery and offspring body mass index, overweight and obesity in adult life: a systematic review and meta-analysis.' *PLoS One.* 2014 Feb 26;9(2):e87896. doi: 10.1371/journal.pone.0087896.

31. Turnbaugh, P.J., Ley, R.E., Mahowald, M.A., Magrini, V., Mardis, E.R., Gordon, G.I. 'An obesity-associated gut microbiome with increased capacity for energy harvest.' *Nature* 21 December, 2006; 444:1027-1031.

32. Larsen, N., Vogensen, F.K., Van der Berg, F.W.J., et al. 'Gut microbiota in Human adults with type 2 diabetes differs from non-diabetic adults.' *PloS One* February 5, 2010; 5(2): e9085

33. Le Chatelier, E., Nielsen, T., Qin, J., et al. 'Richness of human gut microbiome correlates with metabolic markers.' *Nature.* 2013 Aug 29;500(7464):541-6. doi: 10.1038/nature12506.

34. de Weerth, C., Fuentes, S., de Vos, W.M. 'Crying in infants: On the possible role of intestinal microbiota in the development of colic.' *Gut Microbes.* 2013 Aug 9;4(5). [Epub ahead of print]

35. Cabrera-Rubio, R., Collado, M.C., Laitinen, K., et al. 'The human milk microbiome changes over lactation and is shaped by maternal weight and mode of delivery.' *Am J Clin Nutr.* 2012 Sep;96(3):544-51. doi: 10.3945/ajcn.112.037382. Epub 2012 Jul 25.

36. Azad, M.B., Konya, T., Maugham, H., et al. 'Gut microbiota of healthy Canadian infants: profiles by mode of delivery and infant diet at 4 months.' *CMAJ* February 11, 2013 cmaj.

37. Funkhouser, L.J., Bordenstein, S.R. 'Mom knows best: the universality of maternal microbial transmission.' *PLoS Biol.* 2013;11(8):e1001631. doi: 10.1371/journal.pbio.1001631. Epub

2013 Aug 20.

38. Aagaard, K., Ma, J., Antony, K.M., Ganu, R., Petrosino, J., Versalovic, J. 'The placenta harbors a unique microbiome.' *Sci Transl Med.* 2014 May 21;6(237):237ra65. doi: 10.1126/scitranslmed.3008599.

39. Tarazi, C., Agostoni, C., Kim, K.S. 'The placental microbiome and pediatric research.' *Pediatr Res.* 2014 Sep;76(3):218-9. doi: 10.1038/pr.2014.95. Epub 2014 Jul 8.

40. Satokari, R., Gronroos, T., Laitinen, K., et al. 'Bifidobacterium and Lactobacillus DNA in the human placenta.' *Lett Appl Microbiol.* 2009 Jan;48(1):8-12. doi: 10.1111/j.1472-765X.2008.02475.x. Epub 2008 Oct 17.

41. Pert, C.B., Snyder, S.H. 'Opiate receptor: demonstration in nervous tissue.' *Science* 1973;179:1011-4.

42. Csontos, K., Rust, M., et al. 'Elevated plasma beta endorphin levels in pregnant women and their neonates.' *Life Sci.* 1979; 25: 835-44

43. Akil, H., Watson, S.J. et al. 'Beta endorphin immunoreactivity in rat and human blood: Radio-immunoassay, comparative levels and physiological alternatives.' *Life Sci.* 1979; 24: 1659-66

44. Moss, I.R., Conner, H., et al. 'Human beta endorphin-like immunoreactivity in the perinatal/neonatal period.' *J. of Ped.* 1982; 101; 3: 443-46

45. Pedersen, C.S., Prange, J.R. 'Induction of maternal behavior in virgin rats after intracerebroventricular administration of oxytocin.' *Pro. Natl. Acad. Sci.* USA 1979; 76: 6661-65.

46. Olcese, J., Beesley, S. 'Clinical significance of melatonin receptors in the human myometrium.' *Fertil Steril* 2014 Jul 8. pii: S0015-0282(14)00566-4. doi:10.1016/j.fertnstert.2014.06.020. [Epub ahead of print]

47. Sharkey, J.T., Puttaramu, R., Word, R.A., Olcese, J. 'Melatonin synergizes with oxytocin to enhance contractility of human myometrial smooth muscle cells.' *J Clin Endocrinol Metab* 2009 Feb;94(2):421-7. doi: 10.1210/jc.2008-1723. Epub 2008 Nov 11.

48. Hermansson, H., Hoppu, U., Isolauri, E. 'Elective Caesarean Section Is Associated with Low Adiponectin Levels in Cord Blood.' *Neonatology* 2014;105:172-174 (DOI:10.1159/000357178).

49. Varendi, H., Porter, R.H., Winberg, J. 'The effect of labor on olfactory exposure learning within the first postnatal hour.' *Behav Neurosci.* 2002 Apr;116(2):206-11.

50. Tyzio, R., Cossart, R., Khalilov, I., Minlebaev, M., Hubner, C.A., Represa, A., Ben-Ari, Y., Khazipov, R. 'Maternal oxytocin triggers a transient inhibitory switch in GABA signaling in the fetal brain during delivery.' *Science* 2006; 314: 1788-1792.
51. Huxley, J. *Evolution: the modern synthesis.* Allen & Unwin, London, 1942.

CHAPTER 6

1. Barnabei, V.M., Rasmusson, R.L., Bett, G,C. 'Autism and induced labor: is calcium a potential mechanistic link?' *Am J Obstet Gynecol.* 2014 May;210(5):494-5. doi: 10.1016/j.ajog.2014.01.020. Epub 2014 Jan 16.
2. Schlinzig, T., Johansson, S., Gunnar, A., et al. 'Epigenetic modulation at birth – altered DNA-methylation in white blood cells after Caesarean section.' *Acta Paediatr.* 2009;98:1096–9.
3. Almgren, M., Schlinzig, T., Gomez-Cabrero, D., et al. 'Cesarean delivery and hematopoietic stem cell epigenetics in the newborn infant: implications for future health?' *Am J Obstet Gynecol.* 2014 Jun 18. pii: S0002-9378(14)00465-7. doi: 10.1016/j.ajog.2014.05.014.
4. Wong, C.C., Meaburn, E.L., Ronald, A., et al. 'Methylomic analysis of monozygotic twins discordant for autism spectrum disorder and related behavioural traits.' *Mol Psychiatry.* 2014 Apr;19(4):495-503. doi: 10.1038/mp.2013.41. Epub 2013 Apr 23.
5. Gupta, S., Aggarwal, S., Rashanravan, B., Lee, T. 'Th1- and Th2-like cytokines in CD4+ and CD8+ T cells in autism.' *J Neuroimmunol.* 1998 May 1;85(1):106-9.
6. Yonk, L.J., Warren, R.P., Burger, R.A., et al. 'CD4+ helper T cell depression in autism.' *Immunol Lett.* 1990 Sep;25(4):341-5.
7. Becker, K.G. 'Autism, asthma, inflammation, and the hygiene hypothesis.' *Med Hypotheses.* 2007;69(4):731-40. Epub 2007 Apr 6.
8. Brimberg, L., Sadiq, A., Gregersen, P.K., Diamond, B. 'Brain-reactive IgG correlates with autoimmunity in mothers of a child with an autism spectrum disorder.' *Molecular Psychiatry,* (20 August 2013) doi:10.1038/mp.2013.101.
9. Román, G.C., Ghassabian, A., Bongers-Schokking, J.J., et al. 'Association of gestational maternal hypothyroxinemia and increased autism risk.' *Ann Neurol.* 2013 Aug 13. doi: 10.1002/ana.23976. [Epub ahead of print]

10. Modahl, C., Green, L., et al. 'Plasma oxytocin levels in autistic children.' *Biol Psychiatry* 1998; 43 (4): 270-7.
11. Green, L., Fein, D., et al. 'Oxytocin and autistic disorder: alterations in peptides forms'. *Biol Psychiatry* 2001; 50 (8): 609-13.
12. Malek, A., Blann, E., Mattison, D.R. 'Human placental transport of oxytocin.' *J Matern Fetal Med.* 1996 Sep-Oct;5(5):245-55.
13. Saunders, N.R., Habgood, M.D., Dziegielewska, K.M. 'Barrier mechanisms in the brain, II. immature brain.' *Clin. Exp. Pharmacol. Physiol.* 1999;26(2):85-91
14. Braun, L.D., Cornford, E.M., Oldendorf, W.H. 'Newborn rabbit blood-brain barrier is selectively permeable and differs substantially from the adult.' *J. Neurochem* 1980; 34: 147-152.
15. Cornford, E.M., Braun, L.D., Oldendorf, W.H. 'Developmental modulations of blood-brain barrier permeability as an indicator of changing nutritional requirements in the brain.' *Pediatr. Res* 1982;16:324-328
16. Brenton, D.P., Gardiner, R.M., 'Transport of L-phenylalanine and related amino acids at the ovine blood-brain barrier.' *J. Physiol* 1988;402: 497-514
17. Frank, H.J., Jankovic-Vokes, T.,. Pardridge, W.M., Morris, W.L., 'Enhanced insulin binding to blood-brain barrier in vivo and to brain microvessels in vitro in newborn rabbits.' *Diabetes* 1985; 34:728-733
18. Noseworthy, M., Bray, T. 'Effect of oxidative stress on brain damage detected by MRI and in vivo 31P-NMR.' *Free Rad. Biol. Med.* 1998;24:942-951
19. Agnagnostakis, D., Messaritakis, J., Damianos, D., Mandyla, H. 'Blood-brain barrier permeability in healthy infected and stressed neonates.' *J. Pediatr.* 1992;121:291-294.
20. Noseworthy, M., Bray, T. 'Zinc deficiency execerbates loss in blood-brain barrier integrity induced by hyperoxia measured by dynamic MRI.' *PSEBM.* 2000;231:175-182.
21. Schneid-Kofman, N., Silberstein, T., Saphier, O., Shai, I., Tavor, D., Burg, A. 'Labor augmentation with oxytocin decreases glutathione level.' *Obstet Gynecol Int.* 2009;2009:807659. Epub 2009 Apr 16.
22. Modahl, C., Green, L., et al. 'Plasma oxytocin levels in autistic children.' *Biol Psychiatry* 1998; 43 (4): 270-7.
23. Green, L., Fein, D., et al. 'Oxytocin and autistic disorder: alterations in peptides forms.' *Biol Psychiatry* 2001; 50 (8): 609-

13.

24. Demitrack, M.A., Lesem, M.D., Listwak, S.J., et al. 'CSF oxytocin in anorexia.nervosa and bulimia nervosa: clinical and pathophysiologic considerations.' *Am J Psychiatry* 1990 Jul;147(7):882-86

25. Ravelli, G.P., Stein, Z.A., Susser, M.W. 'Obesity in young men after famine exposure in utero and early infancy.' *N Engl J Med.* 1976 Aug 12;295(7):349-53

26. Ravelli, A.C., van der Meulen, J.H., Osmond, C., et al. 'Obesity at the age of 50 y in men and women exposed to famine prenatally.' *Am J Clin Nutr* 1999; 70(5): 811-16

27. Ravelli, A.C.J., van der Meulen, J.H., Michels, R.P. 'Glucose tolerance in adults after prenatal exposure to famine.' *Lancet* 1998; 351: 173-77

28. Stanner, S.A., Bulmer, K., Andres, C. 'Does malnutrition in utero determine diabetes and coronary heart disease in adulthood? Results from the Leningrad siege study, a cross-sectional study.' *BMJ* 1997; 315: 1342-49

29. Power, C., Jefferis, B.J. 'Fetal environment and subsequent obesity: a study of maternal smoking.' *Int J Epidemiol* 2002; 31(2): 413-19

30. Reilly, J.J., Armstrong, J., Dorosty, A.R., et al. 'Early risk factors for obesity in childhood: cohort study.' *BMJ* 2005;330:1357-9.

31. Dalziel, S.R., Walker, N.K., Parag, V., et al. 'Cardiovascular risk factors after antenatal exposure to betamethasone: 30 year follow-up of randomized controlled trial.' *Lancet* 2005; 365: 1856-62

32. Burke, V., Beiling, L.J., Simmer, K., et al. 'Breastfeeding and Overweight: Longitudinal Analysis in an Australian Birth Cohort.' *J Ped* 2005;147: 56-61.

33. Harder, T., Bergmann, R., Kallischnigg, G., et al. 'Duration of Breastfeeding and Risk of Overweight: A Meta-Analysis.' *Am. J. Epidemiol* 2005 Sep1;162(5):397-403.

34. Li, L., Parsons, T.J., Power, C. 'Breastfeeding and obesity in childhood: cross sectional study.' *BMJ* 2003; 327(7420): 904-5

35. Stettler, N., Stellings, V.A., Troxel, A.B., et al. 'Weight gain in the first week of life and overweight in adulthood: a cohort study of European American subjects fed infant formula.' *Circulation* 2005; 111(15):1897-1903

36. Rodekamp, E., Harder, T., Kohlhoff, R., et al. 'Long-term impact of breast-feeding on body weight and glucose tolerance in

children of diabetic mothers: role of late neonatal period and early infancy.' *Diabetes Care* 2005; 28(6):1457-62

37. Goldani, H.A., Bettiol, H., Barbieri, M.A., et al. 'Cesarean delivery is associated with an increased risk of obesity in adulthood in a Brazilian birth cohort study.' *Am J Clin Nutr.* 2011 Jun;93(6):1344-7. doi: 10.3945/ajcn.110.010033. Epub 2011 Apr 20.

38. Barros, F.C., Matijasevich, A., Hallal, P.C., et al. 'Cesarean section and risk of obesity in childhood, adolescence, and early adulthood: evidence from 3 Brazilian birth cohorts.' *Am J Clin Nutr.* 2012 Feb;95(2):465-70. doi: 10.3945/ajcn.111.026401. Epub 2012 Jan 11.

39. Pei, Z., Heinrich, J., Fuertes, E., et al. 'Cesarean delivery and risk of childhood obesity.' *J Pediatr.* 2014 May;164(5):1068-1073.e2. doi: 10.1016/j.jpeds.2013.12.044. Epub 2014 Feb 5.

40. Wang, L., Alamian, A., et al. 'Cesarean section and the risk of overweight in grade 6 children.' *Eur J Pediatr* 2013 Oct;172(10):1341-7. doi: 10.1007/s00431-013-2043-2. Epub 2013 May 25.

41. Blustein, J., Attina, T., Liu, M., et al. 'Association of caesarean delivery with child adiposity from age 6 weeks to 15 years.' *Int J Obes* (Lond). 2013 Jul;37(7):900-6. doi: 10.1038/ijo.2013.49. Epub 2013 Apr 8.

42. Huh, S.Y., Rifas-Shiman, S.L., Zera, C.A. et al. 'Delivery by caesarean section and risk of obesity in preschool age children: a prospective cohort study.' *Arch Dis Child.* 2012 Jul;97(7):610-6. Epub 2012 May 23.

43. Mazaki-Tovi, S., Kanety, H., Pariente, C., etc. 'Determining the source of fetal adiponectin.' *J Reprod Med.* 2007 Sep;52(9):774-8.

44. Hermansson, H., Hoppu, U., Isolauri, E. 'Elective Caesarean Section Is Associated with Low Adiponectin Levels in Cord Blood.' *Neonatology* 2014;105:172-174 (DOI:10.1159/000357178)

45. Ley, S.H., Hanley, A.J., Sermer, M., et al. 'Associations of prenatal metabolic abnormalities with insulin and adiponectin concentrations in human milk.' *Am J Clin Nutr.* 2012 Apr;95(4):867-74. doi: 10.3945/ajcn.111.028431. Epub 2012 Feb 29.

46. Takagi, K., Legrand, R., Asakawa, A., et al. 'Anti-ghrelin immunoglobulins modulate ghrelin stability and its orexigenic effect in obese mice and humans.' *Nat Commun.* 2013;4:2685.

137

doi: 10.1038/ncomms3685.

47. Bellone, S., Rapa, A., Vivenza, D., et al. 'Circulating ghrelin levels in newborns are not associated to gender, body weight and hormonal parameters but depend on the type of delivery.' *J Endocrinol Invest.* 2003 Apr;26(4):RC9-11.

48. Calay, E.S., Hotamisligil, G.S. 'Turning off the inflammatory, but not the metabolic, flames.' *Nat Med.* 2013 Mar;19(3):265-7. doi: 10.1038/nm.3114.

49. Turnbaugh, P.J., Ley, R.E., Mahowald, M.A., et al. 'An obesity-associated gut microbiome with increased capacity for energy harvest.' *Nature* 21 December, 2005; 444:1027-1031.

CHAPTER 7

1. McClellan, M.S., Cabianca, W.A. 'Effects of early mother-infant contact following cesarean birth.' *Obstet Gynecol.* 1980 Jul;56(1):52-5.

2. Smith, J., Plaat, F., Fisk, N.M. 'The natural caesarean: a woman-centred technique.' *BJOG.* 2008 Jul;115(8):1037-42; discussion 1042. doi: 10.1111/j.1471-0528.2008.01777.x.

3. Magee, S.R., Battle, C., et al. 'Promotion of Family-Centered Birth With Gentle Cesarean Delivery.' *J Am Board Fam Med* September-October 2014 vol. 27 no. 5 690-693.

4. Bagci, S., Berner, A.L., et al. 'Melatonin concentration in umbilical cord blood depends on mode of delivery.' *Early Human Development.* Volume 88, Issue 6, Pages 369-373, June 2012

5. Gouchon, S., Gregori, D., Picotto, A., et al. 'Skin-to-skin contact after cesarean delivery: an experimental study.' *Nurs Res.* 2010 Mar-Apr;59(2):78-84. doi: 10.1097/NNR.0b013e3181d1a8bc

6. Kuitunen, M., Kukkonen, K., Juntunen-Backman, K., et al. 'Probiotics prevent IgE-associated allergy until age 5 years in cesarean-delivered children but not in the total cohort.' *J Allergy Clin Immunol.* 2009 Feb;123(2):335-41. doi: 10.1016/j.jaci.2008.11.019. Epub 2009 Jan 8.

7. Jakobsson, H.E., Abrahamsson, T.R., Jenmalm, M.C., et al. 'Decreased gut microbiota diversity, delayed Bacteroidetes colonisation and reduced Th1 responses in infants delivered by Caesarean section.' *Gut.* 2013 Aug 7. doi: 10.1136/gutjnl-2012-303249. [Epub ahead of print]

8. Koyama, A., Toida, I., Nakata, S. 'Osteitis as a complication of BCG vaccination.' *Kekkaku* 2009 Mar;84(3):125-32.

9. Kristensen, I., Aaby, P., Jensen, H. 'Routine vaccinations and child survival: follow up study in Guinea-Bissau, West Africa.' *BMJ*. 2000 Dec 9;321(7274):1435-8.
10. Shann, F. 'Commentary: BCG vaccination halves neonatal mortality.' *Pediatr Infect Dis J* 2012;31:308–309.
11. Aronson, N.E., Santosham, M., Comstock, G.W., et al. 'Long-term Efficacy of BCG Vaccine in American Indians and Alaska Natives.' *JAMA*. 2004;291(17):2086-2091. doi:10.1001/jama.291.17.2086.
12. Odent, M. 'Future of BCG'. *Lancet* 1999; 354: 2170.

CHAPTER 8
1. Odent, M. *Primal Health*, Century Hutchinson, London, 1986.
2. www.primalhealthresearch.com

CHAPTER 9
1. Klaus, M.H., Kennell, J.H. *Maternal-infant bonding*, CV Mosby, St Louis, 1976
2. De Chateau, P., Wiberg, B. 'Long-term effect on mother-infant behavior of extra contact during the first hour postpartum. I. First observations at 36 hours.' *Acta Paediatrica Scand* 1977;66:137.
3. De Chateau, P., Wiberg, B. 'Long-term effect on mother-infant behavior of extra contact during the first hour postpartum. II. Follow-up at three months.' *Acta Paediatrica Scand* 1977;66:145.
4. Schaller, J., Carlsson, S.G., Larsson, K. 'Effects of extended post-partum mother-child contact on the mother's behavior during nursing.' *Infant Behavior and Development* 1979 (2):319-324
5. Terkel, J., Rosenblatt, J.S. 'Humoral factors underlying maternal behaviour at parturition: cross transfusion between freely moving rats.' *J Comp Physiol Psychol* 1972;80: 365-371.
6. Siegel, H.I., Greenwald, M.S. 'Effects of mother-litter separation on later maternal responsiveness in the hamster.' *Physiol Behav* 1978;21:147-149.
7. Siegel, H.I., Rosenblatt, J.S. 'Estrogen-induced maternal behaviour in hysterectomized-ovariectomized virgin rats'. *Physiol Behav* 1975;465-471.
8. Jelliffe, D.B., Jelliffe, E.F.P. (eds). 'The uniqueness of human milk.' *Am J Clin Nutr* 1971;24:968-1009.
9. Jelliffe, D.B., Jeliffe, E.F.P. *Human milk in the modern world*, Oxford University Press, 1978

10. McClelland, D.B., McGrath, J., Samson, R.R. 'Antimicrobial factors in human milk. Studies of concentration and transfer to the infant during the early stages of lactation.' *Acta Paediatr Scand Suppl.* 1978;(271):1-20.
11. Odent, M. 'The early expression of the rooting reflex.' *Proceedings of the 5th International Congress of Psychosomatic Obstetrics and Gynaecology, Rome 1977.* London, Academic Press, 1977: 1117-19.
12. Odent, M. 'L'expression précoce du réflexe de fouissement.' *Les cahiers du nouveau-né* 1978; 1-2: 169-185
13. Virella, G., Silveira Nunes, M.A., Tamagnini, G. 'Placental transfer of human IgG subclasses.' *Clin Exp Immunol.* 1972 Mar;10(3):475-8.
14. Pitcher-Wilmott, R.W., Hindocha, P., Wood, C.B. 'The placental transfer of IgG subclasses in human pregnancy.' *Clin Exp Immunol.* 1980 Aug;41(2):303-8.

CHAPTER 10
1. Laborit, H. '4-hydroxybutyrate.' *Int J Neuropharmacol* 1964;32: 433-451
2. Snead, O.C., Gibson, M. 'Gamma-hydroxybutyric acid.' *NEJM* 2005; 352:2721-2732.
3. Odent, M. 'Champagne and the fetus ejection reflex.' *Midwifery Today* 2003; 65:9.
4. Santhakumar, V., Wallner, M., Otis, T.S. 'Ethanol acts directly on extrasynaptic subtypes of GABA receptors to increase tonic inhibition.' *Alcohol* 2007; 41 (3): 211–21.
5. Liu, S.K., Fitzgerald, P.B., Daigle, M., et al. 'The relationship between cortical inhibition, antipsychotic treatment, and the symptoms of schizophrenia.' *Biol Psychiatry.* 2009 Mar 15;65(6):503-9. Epub 2008 Oct 31.
6. Fattore, L., Cossu, G., Spano, M.S., et al. 'Cannabinoids and reward: interactions with the opioid system.' *Crit Rev Neurobiol.* 2004;16(1-2):147-58.
7. Odent, M. 'Birth under water.' *Lancet* 1983: pp1476-77.

CHAPTER 11
1. Csontos, K., Rust, M., et al. 'Elevated plasma beta endorphin levels in pregnant women and their neonates.' *Life Sci.* 1979; 25: 835-44
2. Akil, H., Watson, S.J. et al. 'Beta endorphin immunoreactivity in

rat and human blood: Radio-immunoassay, comparative levels and physiological alternatives.' *Life Sci.* 1979; 24: 1659-66

3. Moss, I.R., Conner, H., et al. 'Human beta endorphin-like immunoreactivity in the perinatal/neonatal period' *J. of Ped.* 1982; 101; 3: 443-46

4. Kimball, C.D., Chang, C.M., et al. 'Immunoreactive endorphin peptides and prolactin in umbilical vein and maternal blood.' *Am. J. Obstet. Gynecol.* 1987; 14: 104-105

5. Rivier, C., Vale, W., Ling, N., Brown, M., Guillemin, R. 'Stimulation in vivo of the secretion of prolactin and growth hormone by beta-endorphin.' *Endocrinology* 1977 ; 100: 238-41

6. Marais, E.N. *The soul of the white ant, with a biographical note by his son*, Methuen, London, 1937

7. Krehbiel, D., Poindron, P., et al. 'Peridural anaesthesia disturbs maternal behaviour in primiporous and multiporous parturient ewes.' *Physiology and behavior.* 1987; 40: 463-72.

8. Melzack, R. 'Evolution of the neuromatrix theory of pain.' The Prithvi Raj Lecture: presented at the third World Congress of World Institute of Pain, Barcelona 2004. *Pain Pract.* 2005 Jun;5(2):85-94.

9. Evans, S.L., Dal Monte, O., et al. 'Intranasal oxytocin effects on social cognition: a critique.' *Brain Res.* 2014 Sep 11;1580:69-77. doi: 10.1016/j.brainres.2013.11.008. Epub 2013 Nov 14.

10. Melzack, R., Wall, P.D. 'Pain mechanisms: a new theory.' *Science.* 1965 Nov 19;150(699):971–979.

11. Odent, M. 'La reflexotherapie lombaire. Efficacité dans le traitement de la colique néphrétique et en analgésie obstétricale.' *La Nouvelle Presse Medicale* 1975; 4 (3):188

12. Huntley, A.L., Coon. J.T., Ernst. E. 'Complementary and alternative therapies for labor pain: a systematic review.' *Am J Obstet Gynecol* 2004 Jul;191(1):36-44

13. Odent, M. 'Birth under water.' *Lancet* 1983. DOI: http://dx.doi.org/10.1016/S0140-6736(83)90816-4

CHAPTER 12

1. Dorion Sagan and Lynn Margulis, *Origins of sex: three billion years of genetic recombination,* Yale University Press, 1986.

2. Armory Lovins, *Reinventing Fire: Bold Business Solutions for the New Energy Era,* Chelsea Green Publishing Company, 2011.

3. www.worldenergyoutlook.org

4. Odent, M., 'Knitting needles, cameras and electronic fetal

monitors.' *Midwifery Today* 1996; 37: 14-15.

5. Odent, M. 'Knitting midwives for drugless childbirth.' *Midwifery Today* 2004;71: 21-22.

6. McDonald, P. *Oxford Dictionary of Medical Quotations*, p78. Oxford University Press, 2004.

CHAPTER 13

1. Downes, K.L., Hinkle, S.N., Sjaarda, L.A., et al. 'Prior Prelabor or Intrapartum Cesarean Delivery and Risk of Placenta Previa' *American Journal of Obstetrics and Gynecology* 2015 http://www.ajog.org/article/S0002-9378(15)00005-8

2. Levine, L.D., Sammel, M.D., Hirshberg, A., et al. 'Does stage of labor at time of cesarean affect risk of subsequent preterm birth?' *Am J Obstet Gynecol.* 2014 Sep 30. pii: S0002-9378(14)01020-5. doi: 10.1016/j.ajog.2014.09.035. [Epub ahead of print]

3. Odent, M. 'Birth under water.' *Lancet* 1983. DOI: http://dx.doi.org/10.1016/S0140-6736(83)90816-4

4. Odent, M. *The Caesarean,* Free Association Books, London, 2004.

5. Kennell, J., Klaus, M., et al. 'Continuous emotional support during labor in a US hospital.' *JAMA* 1991; 265: 2197-2201.

ADDENDUM

1. Louis Gallien *La selection animale* Presses Universitaires de France, Paris, 1946.

2. Haldane, J., 'Lysenko and Genetics.' *Science and Society* 1940; 4(4).

3. Odent, M. *Childbirth and the Evolution of Homo sapiens* 2nd edition, Pinter & Martin, London, 2014.

INDEX

!Kung San 15, 17

adiponectin 31, 55, 67
aggression, potential for 13
allergic diseases 38, 51
Alzheimer's disease 46
animal experiment 32, 45, 99
antibiotics 31, 51, 72, 117, 120
asthma 38, 51
atopic diseases 51
autism 34, 37, 39-43, 59-64, 66, 69
auto-immune diseases 51
awareness 34, 35-36, 69, 70, 111
ayahuasca 93

bacteriology 34, 47, 49-53, 85
BCG 74-75
beliefs, perinatal 10
birthing pool 96, 114
blood brain barrier 63
brain selective nutrient 18

caesarean *see* pre-labour caesarean *and* in-labour caesarean
cannabis 93
champagne 92, 94
coach 10-11, 20, 96, 112, 115
colostrum 85, 88
cultural conditioning 10, 13, 15, 21, 58, 81, 83, 84, 88-89, 95-97, 98, 106, 112

DHA *see* docosahexaenoic acid
diabetes 38, 51, 52, 75, 118

DNA methylation 48, 61
docosahexaenoic acid (DHA) 19
domination of nature 98, 100, 104-105, 115
dysgenic effects 119, 122
Eipos 16, 17
epidural anaesthesia 43-44, 99
epigenetics 27, 34, 47-48, 60
eugenic effects 122
Everett, Daniel 13-14, 15
evolutionary biology 34, 47, 56-58

fathers 21, 58, 73, 87, 96, 109
fetus ejection reflex 16, 20-21, 73, 92

GABA 32, 55, 61, 91-92
gate control theory of pain 101-103
Gene-environment interaction 39, 41, 63
ghrelin 68
gut flora 31-32, 49, 50, 52-53, 68

homo ludens 77-82
husband-coached childbirth 96

IgA 85
IgG 50, 73, 85
immune system 49-51, 61, 68, 72, 74, 75, 109
in-labour caesarean 26, 29, 31, 45, 48, 52, 67, 70, 113, 114
informational substances 47, 53-56, 67, 79

interdisciplinarity 44, 69
intracutaneous injection of
    sterile water 101-102
intranasal oxytocin 100
iodine 17-18
labour pain 98-103
lumbar reflexotherapy 102

mechanical difficulties 9
melatonin 32, 55, 72, 92, 108
microbiome 31, 49-53, 56, 61,
    68, 75, 87, 106
milk microbiome 31, 50, 52
mouth microbiome 50

natural selection 10, 23, 27, 36,
    116, 117-120
nature, domination of 98, 100,
    104-105, 115
neocortex 20, 22, 89, 90, 91,
    100, 108-109, 112
neocortical control 20, 90, 98,
    100
neocortical inhibition 22, 90,
    91-92, 94, 98, 112
neolithic revolution 13, 57, 100,
    115
noradrenaline 30, 55
normality 23-28, 33

obesity 31, 38, 48, 52, 64-68, 69
oxytocin, synthetic 26-27, 38,
    44, 45, 62-63, 81, 113

pain, gate control theory of
    101-103
paleolithic societies 14, 16
paradigm shift 37, 95, 100, 105,
    113-115
perinatal beliefs 10

Pirahãs 13-14, 17
placenta 27, 33, 42, 50, 62, 67,
    72, 85, 95, 115
placental microbiome 53
potential for aggression 13
pre-labour caesarean 26, 29,
    31-33, 40, 45, 46, 48, 52, 55,
    67-68, 72, 81, 89, 107, 113
pre-midwifery societies 13-17
primal health research 35-46,
    47-48, 49, 51, 54-55, 56, 58,
    59, 64, 69, 77- 82
probiotics 74
psychedelic drugs 92

rituals 10, 14, 95
rooming-in 86

Shiefenhövel, Wulf 16
Shostak, Marjorie 15
skin microbiome 50
skin-to-skin contact 71-72, 75,
    84
support 10, 11, 20-21, 95
symbiosis 104-106
symbiotic revolution 104-110,
    111
synthetic oxytocin 26-27, 38,
    44, 45, 62-63, 81, 113

transgenerational effects 45, 48

uncoupled protein 2 32, 55
undernutrition 45
utopia 106-107, 109-110, 115,
    121-122

vaccination 34, 43
vaginal flora 73
venus figurines 57